TEACH YOURSELF BOOKS

NEW MATHEMATICS

This book is written for those who wish to understand something of the groundwork and purpose of modern mathematics. It introduces the student to real and complex numbers, examines vectors, and discusses number systems and properties, sets and logic, matrices and finally probability. The subject matter is developed logically and presupposes that the reader is only familiar with simple algebra, geometry, trigonometry and the basic processes of arithmetic.

This book forms a useful guide to the principles of complex numbers, vectors, scales of notation and set theory. Building on these, chapters on logical circuits and matrices indicate ways in which the so-called "new" mathematics can be applied.

The Times Educational Supplement

TEACH YOURSELF BOOKS

NEW MATHEMATICS

L. C. Pascoe, M.A.
Headmaster, Ducie High School, Manchester
Formerly Chief Examiner in Mathematics
to various University Boards
Sometime Open Major Scholar in Mathematics,
Hertford College, Oxford

TEACH YOURSELF BOOKS
HODDER & STOUGHTON

ST. PAUL'S HOUSE WARWICK LANE
LONDON EC4P 4AH

First printed 1970
Second edition 1971
Fourth impression 1975

ISBN 0 340 05971 0

Filmset by Keyspools Ltd, Golborne, Lancashire, and Printed and Bound in Great Britain for Teach Yourself Books, Hodder and Stoughton by Richard Clay (The Chaucer Press), Ltd., Bungay, Suffolk

To
WINSOME, HILARY and PETER

INTRODUCTION

THE CONCEPT of modern mathematics has crystallized largely as a result of the invention, in Britain, of the electronic digital computer, with its need for Boolean algebra—a branch of mathematics which had lain neglected for a century. The discovery gave impetus, among other things, to mathematical developments in schools and colleges of the United States of America, when there had already existed unrest about the low standard of mathematical attainment. It was not long before we, in this country, began to follow suit in our thinking, although influential circles in Britain were slow at first to realize the tremendous impact the new logical machines would have on our way of life. The present author referred to this as early as 1958, in *Teach Yourself Arithmetic* (Chapter 17).

Teach Yourself Mathematics is written for those who wish to understand something of the groundwork and purpose of modern mathematics. This book is especially designed for the student who wishes to pursue his studies independently of tuition. The subject-matter is developed logically and presupposes that the student is familiar only with simple algebra (as far as the solution of simultaneous linear equations and very easy quadratic equations), geometry (elementary properties of triangles and parallelograms), trigonometry (the definitions of the ratios sine, cosine and tangent) and the basic processes of arithmetic. After the introduction of numbers, real and complex, vectors are examined. Then follow number systems and properties, sets and logic (including circuit theory). Matrices are covered without reference to their vector properties and, finally, probability is introduced, firstly as an independent topic and then linked with set theory. Chapters 1–8 are of a simple and straightforward nature. Chapters 9 and 10 require thought and the reader will find it helpful to have paper and pen to hand when dealing with circuit theory. Chapters 11 and 12 can be understood fully without reference to chapters 9 and 10.

In writing a book of this nature decisions have to be taken, on the sheer question of length, as to subject-matter to be included or omitted. It was decided to exclude flow diagrams, programming for computers and a discourse on the actual electronic construction of the machines themselves. There is good reason for the decision. Those few people who will actually operate the machines will have special tuition. Two other omissions should be mentioned, namely topology and functions. The former is an advanced study and, at a level which would have a place in such a work as the present one, would be pointless to include. Functions (and mapping) could have a stronger case made for discussion, but they were not essential for the present purpose.

It is worthy of mention that *A Parent's Guide to the New Mathematics*, by Evelyn Sharp, edited by L. C. Pascoe, and published by the English Universities Press Ltd., is a very easy and lighthearted introduction to the subject. Some students may well find it helpful firstly to read this in order to familiarize themselves with the principles before studying the much more meaty contents of *Teach Yourself New Mathematics*.

L. C. PASCOE

Ducie High School,
Manchester, 1968

INTRODUCTION TO THE SECOND EDITION

The only significant alteration to the first edition, which has proved to be necessary, has been the reconstruction of a part of Chapter 8. Apart from this, a number of minor adjustments and corrections have been effected.

1971   L.C.P.

CONTENTS

INTRODUCTION

CHAPTER 1

REAL NUMBERS

Section	Page	Section	Page
1. The Origination of our Number System	13	6. Proof that $\sqrt{2}$ is Irrational	22
2. The Development of Numbers	18	7. The Bounding of Irrationals by Rationals	23
3. The Number Line	19	8. Approximations to Square Roots	24
4. Rational Numbers	20	9. The Set of Real Numbers	24
5. Irrational Numbers	21		

CHAPTER 2

COMPLEX NUMBERS

1. Imaginary Numbers	26	3. The Four Rules applied to Complex Numbers	30
2. Complex Numbers	29	4. Cartesian Ordered Pairs	33

CHAPTER 3

VECTORS

1. Scalars and Vectors	36	5. Abbreviated Notation	46
2. Coordinates and Vectors	37	6. Applications of Vectors to Problems on Velocity	50
3. The Addition of Vectors	43	7. Localised Vectors	55
4. The Multiplication of a Vector by a Scalar	45		

CHAPTER 4

SCALES OF NOTATION

1. Number Systems 60
2. The Seximal Scale 61
3. The Binary Scale 63
4. Bicimals 68
5. The Octal Scale 73

CHAPTER 5

MODULAR ARITHMETIC

1. Modulo Systems 76
2. Algebraic Congruences .. 79
3. Tests for Factors of Whole Numbers 81

CHAPTER 6

INTRODUCTION TO SETS

1. Sets 86
2. Subsets 87
3. Symbolism 89
4. Venn Diagrams 90
5. Intersection of Sets 92
6. Union of Sets 95

CHAPTER 7

INEQUALITIES

1. Inequalities 101
2. Set Notation for Inequalities 102
3. Axioms of Inequalities . 104
4. Graphical Representation of Inequalities involving two Variables 107
5. Cartesian Product 111

CHAPTER 8

SET THEORY

1. Standard Formulae 115
2. Intersection and Union of three Sets 116
3. Highest Common Factor and Lowest Common Multiple 117
4. The Complement of a Set 119
5. Symmetric Difference .. 125
6. Abbreviated Notation .. 127

CHAPTER 9

MATHEMATICAL SYMBOLIC LOGIC

1. Logic 130
2. Notation 133
3. Truth Tables 136
4. Standard Results 139
5. The NOR Function 141

CHAPTER 10
LOGICAL CIRCUIT THEORY

1. Abbreviated Notation for Symbolic Logic 144
2. AND and OR Circuits 145
3. The NOR Circuit 150
4. Logical Design: the Binary Half Adder 153
5. The Whole Adder 156
6. The NOR-gate Half Adder and Whole Adder 158
7. Binary Addition Computer 163

CHAPTER 11
MATRICES

1. Matrices 164
2. Simple Matrix Properties 166
3. Multiplication of Matrices by Real Numbers 169
4. Multiplication of Matrices together 171
5. Matrix Method of Solution of Simultaneous Linear Equations 176
6. Power of a Matrix 180
7. Application of Matrices to Public Opinion Polls 183

CHAPTER 12
PROBABILITY

1. Probability for a Single Event 187
2. Exclusive Events 192
3. Independent Events 194
4. Interdependent Events 196
5. Binomial Distribution 199
6. Permutations and Combinations 201
7. Sets and Probability 205

ANSWERS

CHAPTER 1

REAL NUMBERS

1. The Origination of our Number System

Mathematics has rightly been called the queen and servant of science. As new developments in abstract mathematics have appeared so have attempts to convert these to practical scientific use and, conversely, as the needs of science have changed so has the emphasis on aspects of mathematics. Today, the greatest single impact on such progress has been the development of the electronic computer.

Mathematics has not progressed historically as a steady maturation of logical thought, but by irregular and intermittent steps, sometimes with centuries intervening between notable landmarks of advancement. The subject is essentially a study requiring men of vision to outline new processes. Such men are to be found only at infrequent intervals.

In the very earliest times only the most primitive ideas of arithmetic were required. A man would not have needed to count, in prehistory, but he would have known his possessions by their individuality. Once, however, a system of barter arose—with the beginnings of civilization—it was necessary to learn to assess, say, how many sheep were being exchanged for a specific number of pottery utensils. Perhaps a certain number of stones was placed in a heap, to represent the value of a sheep, and another number of stones, in a separate pile, represented the value of a cooking-pot. At the end of a transaction involving several sheep and utensils, the stones could have been taken one at a time from each heap (a one-to-one correspondence)[1] and any surplus stones left in one heap, when the other was exhausted, could have been balanced by negotiation—maybe the acquisition of one more small pot. Whether or not this

[1] An important concept in modern geometry.

system was actually adopted is immaterial. The example is intended to illustrate the idea of using stones as *place-holders* for units of value, and it would only have needed a matter of time to have evolved abstract counting (one, two, three, etc.) instead. Thus we have the development of the *natural numbers* (i.e. 1, 2, 3 . . .), otherwise called *positive integers*. These then are place-holders for units of any kind, not merely for money or value.

At first, in counting, people used the five fingers of one hand and this gave rise to the seximal scale (or scale of six), which can be read 1, 2, 3, 4, 5, 10, 11, 12, 13, 14, 15, 20, 21 . . ., wherein there are six symbols, including zero. Until very recent years, this scale was widely used in the Orient. Simple abacuses (counting frames), still to be found in China and Japan,[2] are clearly designed for this system.

It is equally easy to see how ten has become an important number, because people possess ten fingers and ten toes. From this characteristic of the human race there developed the denary scale (or scale of ten). There is, however, an odd point about the seximal, denary and other scales. One might have expected, in the *denary* scale, ten single symbols (digits) to represent fingers thus: 1, 2, 3, 4, 5, 6, 7, 8, 9, t followed by 11, 12, . . ., 19, 1 t, and so on, but in fact we only use nine natural digits (1, 2, . . . 9). To these we add zero, which is not natural at all, and thus obtain 1, 2, . . . 9, 10, 11, . . .

We now have, in the scale of *six*, the six symbols 0, 1, 2, 3, 4, 5 and, in the scale of *ten*, the ten symbols 0, 1, 2, 3, 4, 5, 6, 7, 8, 9, so all is well. We take it for granted, but the introduction of zero was a difficult concept. Whereas it was easy to visualize the number 4 as a place-holder for 4 cows, 4 arrows or even 4 wives, it was a much more sophisticated idea to imagine 0 as representing nothing whatsoever, e.g. a man who did not possess a goat was the same as a man who owned 0 goats. Once this subtlety was mastered, progress in numerals was rapid, although for a long time counting was restricted to positive integers.

It is possible that the introduction of zero was a direct outcome of the use of the abacus. Suppose, for example, that we are members of a primitive race with only a rudimentary knowledge of arithmetic and that we add 4 and 3 in the scale of six, wherein we are familiar with only the digits 1, 2, 3, 4, 5. (See Fig. 1.) We have never even heard of zero!

[2] See *Arithmetic* (*Decimalised and Metricated*), by the Author, Teach Yourself Books.

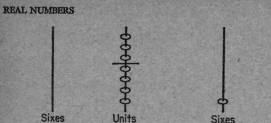

Fig. 1

Transferring six beads from the units column to the sixes column and replacing them by one six, we have one six and one unit, i.e. 11 (in the seximal scale).

Suppose now, however, we add 4 and 2 in this scale. (Fig. 2.)

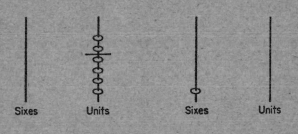

Fig. 2

Transferring six units to the sixes column and replacing them by one six, we have one six and no unit, i.e. 1*. Clearly we need some symbol to replace the asterisk and we adopt 0 to represent zero or nothing at all, so in Fig. 2 we get 10. This leads to scale-of-six counting

1, 2, 3, 4, 5, 10, 11, 12, 13, 14, 15, 20, 21 . . .

as indicated earlier. There are *six* symbols in all, from which arises the name seximal.

In like manner we can evolve counting in the scale of ten, with which we are all so familiar in everyday life:
1, 2, 3, 4, 5, 6, 7, 8, 9, 10, 11, 12, 13, 14, 15, 16, 17, 18, 19, 20 . . .
The members of this system are, as already mentioned above, the natural numbers, or positive integers. The nomenclature logically accords with the theory of numbers, for *whole numbers* (integers) represent whole real-life things, when *positive*, and hence also we can think of them as *natural* in that they are place-holders for natural facts.

Exercise 1

1. At this stage the student may well find it instructive to make a simple abacus. Not only will this help to elucidate the work of this chapter but it will be of value for practice in the four rules (addition, subtraction, multiplication and division) of arithmetic, where numbers are expressed in other scales (e.g. the binary scale and octal scale) later in the book.

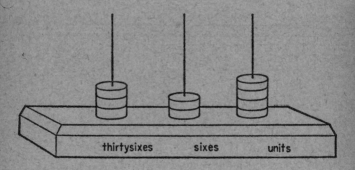

Fig. 3

Figure 3 illustrates an abacus which, in a few minutes, can be constructed at home. Three stiff wires are inserted into a block of wood, the only restriction as to the size of the block being convenience, but 5 in. in length, $1\frac{1}{2}$ in. in width and $\frac{3}{4}$ in. in height are reasonable dimensions. The thickness of the wire depends on the size of the holes in beads which are readily obtainable. Some two dozen beads should be adequate but it is advisable to avoid the use of fiddling little things. The wires need to be long enough to accommodate, say, 12 beads each during working. The above apparatus will, in the seximal scale, give *units, sixes* and *thirty-sixes,* the value of a bead in each column being six times the value of a bead in the preceding column. The largest total, 555, which can therefore be shown on the abacus will become, when converted to the denary scale,

$$5 \times 36 + 5 \times 6 + 5 = 180 + 30 + 5 = 215.$$

This is because in the scale of 6, in which the number 6 does not exist, we cannot finish our calculations with more than 5 beads on any one wire. Thus, for three wires, we cannot finish with more than 15 beads.

The seximal number which is actually shown in Fig. 3 is

$$324 \text{ (seximal scale)} = 3 \times 36 + 2 \times 6 + 4 \text{ (denary scale)}$$
$$= 108 + 12 + 4 \quad (\quad ,, \quad ,, \quad)$$
$$= 124 \quad (\quad ,, \quad ,, \quad)$$

It will be observed that the abacus can be used for adding and subtracting denary numbers just as simply as for numbers in the scale of six, or in fact in any other scale. We merely read the wires, from right to left as, say, *units, tens, hundreds.*

With four wires instead of three, our columns would be as follows:

Scale of six: $6^3, 6^2, 6, 1$
Scale of ten; $10^3, 10^2, 10, 1$

In the seximal scale, the largest number then obtainable, 5555, would then convert to

$$5 \times 6^3 + 5 \times 6^2 + 5 \times 6 + 5 = 1295 \text{ (denary scale)}$$

and 20 beads would be needed.

Similarly, in the denary scale, with four wires the largest number obtainable would be

$$9 \times 10^3 + 9 \times 10^2 + 9 \times 10 + 9 = 9999 \quad \text{(denary scale)}$$

and 36 beads would be needed.

2. Use the abacus with three wires to find the following results, all the numbers being in the seximal scale and the answer being given in the same form:

(a) $5+4$,   (b) $3+5+4$,   (c) $11-4$,   (d) $3+12-5$,
(e) $23+15$,   (f) $45+12+24$, (g) $315-240$, (h) $502-455+110$,
(i) 3×4, (j) 4×22, (k) 2×153, (l) $410\div2$.

3. Convert to the denary scales all the numbers in Qn. 2 above. Convert, also to the same scale, all the answers obtained. Use the abacus in the denary scale to show that the answers are correct.

2. The Development of Numbers

The numerals of the past were often cumbersome. Let us imagine adding MCXLV and DCCVIII in Roman days, if an abacus were not to hand! Various notations gained transient favour but the outstanding achievement was the introduction of the Hindu–Arabic number system, made known in Italy by Leonardo of Pisa (called Fibonacci), whose work was based on an Arabic treatise. We use this presentation today.

Throughout the remainder of the book, numbers are in the *denary* (*decimal*) *system unless specifically stated to be otherwise*. We shall rarely use the word *decimal* unless actually referring to numbers in which the *decimal point* occurs.

After integers were comprehended, fractions were introduced to represent parts of things. Not unnaturally, early ideas were far from elegant. The Romans used only fractions with a denominator of 12, such as $\frac{5}{12}$, $\frac{8}{12}$. The Egyptians were happy with any denominator but used only a numerator of 1, e.g. $\frac{1}{7}$, $\frac{1}{144}$. This must have been a nightmare, for they did not permit the repetition of the same fraction in any one expression. An illustration should make this clear:

$2\div7$ could be written as $\frac{1}{4}+\frac{1}{28}$, but not as
$\frac{2}{7}$ (as we would write it) nor as $\frac{1}{7}+\frac{1}{7}$.

For some reason they allowed $\frac{2}{3}$ as an exception.

Next in the progress of numerals came the invention of negative numbers. Here again was presented to our forebears a problem the magnitude of which we rarely consider nowadays. Suppose a man had 5 horses and acquired 3 more. The process of adding presented no difficulty, for

$$5+3 = 8$$

and therefore the addition of two natural numbers led to another natural number. (Mathematically, we can say that the natural numbers are *closed under addition*.) Suppose now, however, that instead of adding to his collection of 5 horses he disposed of some, say 3, then

$$5-3 = 2$$

and he still had a natural number as the result. The horse-dealer may, on the other hand, have sold more horses than he actually had at the time. A wealthy customer perhaps offered a good price for 7 horses and the dealer knew that he could soon acquire the extra animals needed. He then had the situation in the sale that

$$5-7 = -2,$$

meaning a *debit* of 2 (in this case, the owing of 2 horses). Whatever -2 was, it certainly was not a natural number. We investigate it more carefully anon.

The symbols $+$ and $-$ seem to have been taken in medieval times from marks on sacks of grain, the $+$ (*surplus*, abbreviated to *plus*) indicating that a certain sack was heavier than a standard weight (i.e. had something added to it) and the $-$ (*minus*) that another sack weighed less than a standard weight (i.e. was lacking in quantity).

3. The Number Line

If we draw a *number line*, a horizontal line with the integers marked on it equidistantly in order from left to right, we can see the above processes very clearly. They hardly need exposition.

$$5+3 = 8$$
$$5-3 = 2$$

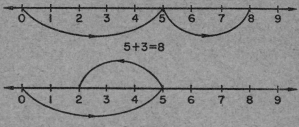

$$5+3=8$$

$$5-3=2$$

Fig. 4

The extension to negative numbers is just as easy if we extend the number line to the left.

$$5 - 7 = -2$$

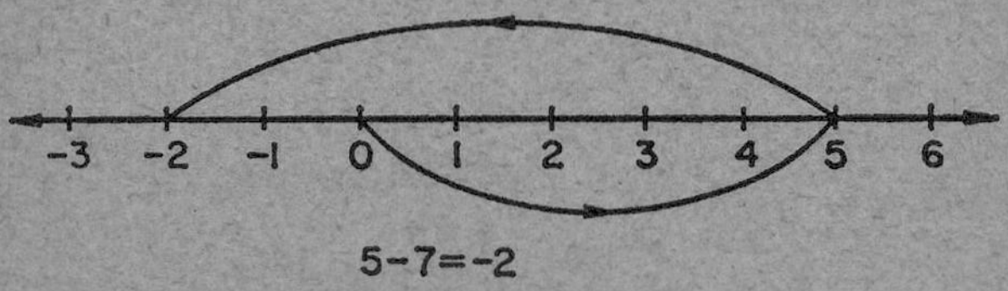

$$5 - 7 = -2$$

Fig. 5

Thus the horse-dealer's activities are illustrated in figs. 4 and 5 above.

It has already been mentioned that the natural numbers are closed under addition, but obviously this is not true of subtraction (for although $5 - 3 = 2$, which is a natural number, we have $5 - 7 = -2$, which is not).

4. Rational Numbers

The whole collection of numbers (positive and negative, integers and fractions, together with zero) so far discussed constitutes the *set* of *rational numbers*. Illustrations of such numbers are

$$17, \quad -4, \quad \tfrac{3}{4}, \quad 0, \quad 0.66, \quad -2\tfrac{3}{7}.$$

The set of rational numbers is closed under addition, subtraction, multiplication and division (if we exclude the single case of division by zero).

It is interesting to observe that all rational numbers can be expressed as recurring decimals (where we include terminating decimals as a special case of recurring decimals) and as fractions such as a/b, where a and b are integers.

For example, $\tfrac{3}{8} = 0.375$ (i.e. $0.375\dot{0}$, where the final zero recurs)

$$3\tfrac{1}{6} = 3.1\dot{6} \text{ (i.e. } 3.1666\ldots)$$
$$\tfrac{1}{7} = 0.\dot{1}4285\dot{7} \text{ (i.e. } 0.14285714285714\ldots)$$

5. Irrational Numbers

Every rational number has a corresponding point on the number line. If all the rational numbers are marked on the number line, they will be so close together that the line will *appear* to be continuous, but it is not so in fact, for scattered among them there is an infinity of points (on the number line) for which there is no corresponding rational number.

Let us consider, in Fig. 6, a square $OABC$ of side 1 unit. Then $OA = 1$. By Pythagoras's theorem

$$OB^2 = OA^2 + AB^2 = 1^2 + 1^2 = 2$$
$$\therefore OB = \sqrt{2}.$$

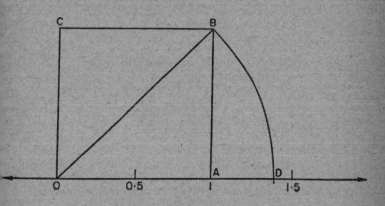

Fig. 6

Draw OA produced as the number line, with O as the point representing zero and A representing the number 1. With centre O and radius OB draw an arc to cut OA produced at D, then $OD = \sqrt{2}$. Now OD lies between 1·4 and 1·5, because $(1·4)^2 = 1·96$ and $(1·5)^2 = 2·25$, whereas $(\sqrt{2})^2 = 2$, which lies between 1·96 and 2·25. More accurately, $\sqrt{2} = 1·4142136 \ldots$, a decimal number which does not recur. Such a number is called *irrational*. There is no *rational* point for it on the number line.

The existence of *surds* (irrational roots of numbers), such as $\sqrt{2}, \sqrt{3}, \sqrt{5}$, was a great worry to the early mathematicians. They tried hard to rationalize them, but without success. Had they had the necessary knowledge, they could have saved themselves much time and trouble, for the proof that $\sqrt{2}$ is irrational is quite straightforward. It is given below but the reader may prefer to omit Section 6 until after studying some later chapters. The continuity of the work is not affected by this temporary omission.

6. Proof that $\sqrt{2}$ is Irrational

We first note that (*a*) the square of an even integer (an integer is a whole number) is itself even (the even number can be written as $2k$, where k is an integer, $\therefore$ its square is $4k^2$, which is certainly divisible by 2), (*b*) if the square root of an even integer is an integer, then it is an even integer. (We introduce the symbol $\Rightarrow$ meaning "implies that". Suppose the given even integer is N and its square root is *odd*, say $2k+1$, where k is an integer, then

$$\sqrt{N} = 2k+1 \Rightarrow N = (2k+1)^2,$$

$$\begin{aligned}
\text{i.e. } N &= 4k^2+4k+1 \\
&= 4(k^2+k)+1 \\
&= \text{Even number}+1 \\
&= \text{Odd number}
\end{aligned}$$

but we are told that N is even, and so we have shown that N is both even and odd, which is absurd $\therefore$ the square root is not odd, i.e. if it is an integer, it is even.)

Let us suppose 2 has a *rational* square root, then this root can be put in the form a/b, where a and b are positive integers having no common factor, any such factor having been removed. (Decimals can be put in this form, of course. For example, $1 \cdot 44 = \frac{144}{100} = \frac{36}{25}$, and we would have $a = 36$, $b = 25$.)

Now

$$\frac{a}{b} = \sqrt{2} \Rightarrow \frac{a^2}{b^2} = 2, \text{ i.e. } a^2 = 2b^2$$

This shows that a^2 is an even integer

$$\Rightarrow a \text{ is an even integer (from } (b) \text{ above)}$$

$$\Rightarrow a = 2k, \text{ where } k \text{ is an integer}$$

but $2b^2 = a^2$, and so $2b^2 = 4k^2$

$$\Rightarrow b^2 = 2k^2$$
$$\Rightarrow b^2 \text{ is an even integer}$$
$$\Rightarrow b \text{ is an even integer}$$

Hence a and b are both even integers and therefore have a common factor 2. This is contrary to the hypothesis that they have no common factor.

$$\therefore \text{ 2 does not have a rational square root.}$$

7. The Bounding of Irrational Numbers by Rationals.

The Greeks discovered a squeezing method of enclosing irrationals between successive pairs of rational fractions which came ever closer together. They constructed a ladder of numbers in the following manner:

a	b
1	1
2	3
5	7
12	17
29	41

The ratio $\dfrac{b}{a}$ for each rung of the ladder, going downwards, is a nearer approximation to $\sqrt{2}$. Furthermore, the approximations lie alternately either side of the required surd ($\sqrt{2} = 1{\cdot}414213\ldots$)

1st rung $\quad \dfrac{b}{a} = \dfrac{1}{1} \ = 1{\cdot}0 \qquad\qquad$ too small

2nd rung $\quad \dfrac{b}{a} = \dfrac{3}{2} \ = 1{\cdot}5 \qquad\qquad$ too large

3rd rung $\quad \dfrac{b}{a} = \dfrac{7}{5} \ = 1.4 \qquad\qquad$ too small

4th rung $\quad \dfrac{b}{a} = \dfrac{17}{12} = 1{\cdot}416\ldots \qquad$ too large

5th rung $\quad \dfrac{b}{a} = \dfrac{41}{29} = 1{\cdot}4131\ldots \qquad$ too small, and so on.

It is clear that each result is nearer to the actual surd than the result before.

The acceptance of irrationals as numbers ranking *pari passu* with rationals is attributed to the work of Plato.

8. Approximations to Square Roots

The foregoing ideas suggest an interesting method of ascertaining square root approximations for surds. Although fundamentally important in its mathematical implications, it does not supersede the usual process for square roots described in standard books of arithmetic, for it would be cumbersome to apply when several significant figures were needed in the result.

The procedure is best illustrated by an example.

Example. Find $\sqrt{5}$.

Let $\sqrt{5} = y$, then $y^2 = 5$,

$$\text{i.e. } y = \frac{5}{y}, \text{ or expressed differently, } y \times \frac{5}{y} = 5.$$

We guess a value for y, say 2, then $5 \div 2 = 2 \cdot 5$.

$$\therefore 2 \times 2 \cdot 5 = 5.$$

Clearly 2 is too small and $2 \cdot 5$ is too big.
Let us try $2 \cdot 2$, then $5 \div 2 \cdot 2 \simeq 2 \cdot 27 \ldots$

$$\therefore 2 \cdot 2 \times 2 \cdot 27 \simeq 5.$$

We now try $2 \cdot 24$, then $5 \div 2 \cdot 24 \simeq 2 \cdot 23 \ldots$

$$\therefore 2 \cdot 23 \times 2 \cdot 24 \simeq 5.$$

We are now getting a fairly close approximation, $2 \cdot 235$.
Actually $\sqrt{5} \simeq 2 \cdot 236$ (where $\simeq$ means "approximately equals").

It will be observed that in this example each new trial number is as near half way between the two numbers obtained on the previous line as can be determined, using only one more decimal place each time.

9. The Set of Real Numbers

The set of rational numbers together with the set of irrational numbers constitute the set of *real* numbers. If all tne real numbers were marked on the number line, it would be continuous and we could, in modern terminology, define the number line as the set of points representing the real numbers.

Exercise 2

1. Given the numbers
$$3, -2, \sqrt{5}, 1\tfrac{1}{3}, -\sqrt{16}, 4\cdot3, 2+\sqrt{7}, 0, \sqrt{-2},$$
state which are (a) natural, (b) rational, (c) real, (d) non-real (i.e. do not lie on the real number line).

2. Show that $\sqrt{3}$ lies between $1\tfrac{1}{2}$ and $1\tfrac{3}{4}$.

3. Find the system by which the rungs of the ladder on page 23 are constructed. Write down the next two rungs and hence work out the next approximation to $\sqrt{2}$ as far as the fourth decimal place.

4. Show that $\sqrt{6}$ lies between $2\cdot4$ and $2\cdot5$. Find, correct to 2 decimal places, the closest approximations to $\sqrt{6}$. (Use the method of Section 6 above.)

5. Show that $\tfrac{1}{9} = 0\cdot\dot{1}$ and that $\tfrac{1}{90} = 0\cdot0\dot{1}$. Hence, express $0\cdot1\dot{2}$ and $0\cdot1\dot{3}$ as proper fractions in lowest terms.

6. Using the method of Section 8, find, correct to 2 decimal places, the square roots of 7, 10 and 17.

CHAPTER 2

COMPLEX NUMBERS

1. Imaginary Numbers

Let us consider the equation $z^2 + 4 = 0$, i.e. $z^2 = -4$. We can see at once that the solution set does not lie on the real number line we considered in Chapter 1, for there is no real number whose square is negative. If we try solutions $z = 2$ or $z = -2$ to the above equation, we have $z^2 = 2^2 = 4$ or $z^2 = (-2)^2 = 4$, and so neither satisfies it. We must, therefore, be seeking a new kind of number.

Solving $z^2 = -4$, we have

$$z = \pm\sqrt{-4} = \pm 2\sqrt{-1} \quad [\text{for } -4 = 4(-1)].$$

If we now write $i = \sqrt{-1}$, then $z = \pm 2i$. We cannot mark points on the real number line to correspond to these answers but suppose that we take a line, through the zero point on the real number line and perpendicular to it. Let us mark points $+2$ and -2 units from zero on this new line and say that they represent $2i$ and $-2i$ respectively. This line we call the *imaginary number line* or *imaginary axis* (for a reason which will soon become clear). At the same time the real number line can be renamed the *real axis*. For convenience we abbreviate these to *Im* and *Rl* axis, on occasion. The zero point we call the *origin*. A graphical picture with these axes is called an *Argand diagram*.

For our representation of $2i$ and $-2i$ to be valid, we need to say that multiplication by i rotates numbers (in this case 2 and -2) through $90°$. We also adopt, for convenience, an anticlockwise sense of rotation. We illustrate the solution set to $z^2 + 4 = 0$ in Fig. 7 below.

Does the process work if we pursue the matter further? Suppose we now multiply our number $2i$ by i, then

$$2i^2 = 2(\sqrt{-1})^2 = 2(-1) = -2.$$

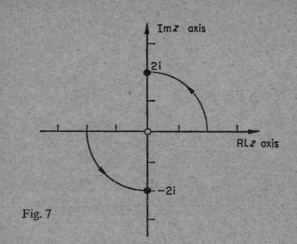

Fig. 7

Now this would mean that 2 has been converted to −2 by operating twice with i, but this clearly fits our suggestion, for a 90° rotation by i applied once would lead to a 180° rotation for i to be applied twice in succession (i.e. for i^2 to be applied) and this would bring 2 into the −2 position on the real axis (Fig. 8).

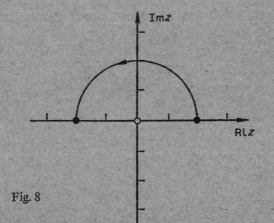

Fig. 8

Continuing thus, $2i^3 = 2i(i^2) = -2i$, would lead to a $270°$ rotation and this again fits our hypothesis. Lastly, a complete rotation of $360°$ corresponds to $2i^4 = 2(i^2)^2 = 2(-1)^2 = 2$, which brings the real number 2 back to its original position on the real number line. Higher powers of i merely lead to repetition, e.g. $2i^{17} = 2i(i^4)^4 = 2i$, for every fourth power of i is $+1$.

Summarizing the powers of i, we have

$$i = i, i^2 = -1, i^3 = -i, i^4 = 1, i^5 = i, i^6 = -1, i^7 = -i, i^8 = 1,...$$

We could, in the above discussion, have chosen any other *positive* number instead of 4 in the equation $z^2 + 4 = 0$. Such a number, being positive, could be called a^2 (where a is real). The solution set of $z^2 + a^2 = 0$ is $z = ai$ or $-ai$.

In set notation, explained in Chapter 8, this can be written

$$\{z : z^2 + a^2 = 0\} \Rightarrow z = ai \text{ or } -ai,$$

which is read "If z is a member of the set of numbers such that $z^2 + a^2 = 0$, then z is one of the two numbers ai or $-ai$."

Example. Simplify the following numbers, stating which are real and which are imaginary: (a) $3i^3$, (b) $(2i)^5$, (c) $(3i)^6$, (d) $5i^{11}$.

We have   (a) $3i^3 = -3i$,   (b) $(2i)^5 = 2^5i^5 = 32i$,
            (c) $(3i)^6 = 3^6i^6 = -729$,   (d) $5i^{11} = 5i^3 = -5i$,

using the method of simplification of powers of i shown above.

Of the given numbers, (a), (b), (d) are imaginary and (c) is real.

Exercise 3

1. Simplify the following numbers, stating which are real and which are imaginary: (a) $4i^3$, (b) i^{10}, (c) $(2i)^7$, (d) $6i^{15}$, (e) $-\frac{1}{2}i^{14}$. Plot the numbers in (a), (b), (d) and (e) on the Rl and Im axes, using a scale of $\frac{1}{2}$ in. to 1 unit.

2. Show that $1/i = -i$ and deduce that dividing a real or imaginary number by i is equivalent to rotating it *clockwise* through $90°$ on an Argand diagram.

3. Simplify   (a) $(3i^3)(2i^2)$,   (b) $(-i)(-2i^4)$,   (c) $(2i)^2(3i^5)$,
(d) $(\frac{1}{2}i^2)^3(2i)^5$,   (e) $5i^{17} - 3i^{15}$,   (f) $i^4 - 2i^6 + 3i^8$.

4. Solve the equations $4z^2 - 9 = 0$ and $4z^2 + 49 = 0$.
Plot the results on an Argand diagram.

2. Complex Numbers

The numbers considered in Section 1 above are represented by points which lie either on the real or on the imaginary axis. Suppose, however, we have a number which is partly real and partly imaginary, say $2+3i$. We would need to proceed 2 units to the right of the origin (along the real axis) and then 3 units upwards (parallel to the imaginary axis), for multiplying 3 by i rotates the direction of travel (of 3) anticlockwise through $90°$.

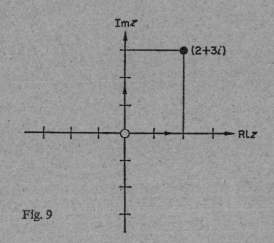

Fig. 9

In Fig. 9 above, we see that if we had taken $3i+2$ we would have reached the same point. Thus

$$3i+2 = 2+3i.$$

We are now in a position to generalize. Any *complex* number z may be put in the form $a+ib$, where a and b are real and $i = \sqrt{-1}$; i.e. we have

$$z = a+ib = ib+a.$$

If $a = 0$, then $z = ib$, which is an imaginary number.
If $b = 0$, then $z = a$, which is a real number.
It follows that real and imaginary numbers are special cases of complex numbers.

3. The Four Rules applied to Complex Numbers

Suppose we have two complex numbers, $z_1 = a + ib$, $z_2 = c + id$.

(a) *Addition.*
$$\begin{aligned} z_1 + z_2 &= a + ib + c + id \\ &= (a + c) + i(b + d) \\ &= p + iq, \text{ say,} \end{aligned}$$

where $p = a + c$, $q = b + d$, are real. Hence the sum of two complex numbers is itself a complex number.

(b) *Subtraction.*
$$\begin{aligned} z_1 - z_2 &= a + ib - (c + id) \\ &= a + ib - c - id \\ &= (a - c) + i(b - d) \\ &= r + is, \end{aligned}$$

where r, s are real. Hence subtraction of a complex number from another leads to a complex number.

(c) *Multiplication.*
$$\begin{aligned} z_1 z_2 &= (a + ib)(c + id) \\ &= ac + ibc + iad + i^2 bd \end{aligned}$$

Now $i^2 = -1$, from Section 7 above,

$$\begin{aligned} \therefore z_1 z_2 &= (ac - bd) + i(bc + ad) \\ &= t + iu \end{aligned}$$

where t, u are real. Hence the product of two complex numbers is a complex number.

(d) *Division*

$$\frac{z_1}{z_2} = \frac{a + ib}{c + id}.$$

Here, we cannot proceed further until we have made the denominator real, if this is possible. Consider

$$\begin{aligned} (c + id)(c - id) &= c^2 - i^2 d^2 \\ &= c^2 + d^2. \end{aligned}$$

We see that, if we multiply the numerator and denominator of the right-hand side by $c - id$, we have

$$\frac{z_1}{z_2} = \left(\frac{a + ib}{c + id}\right)\left(\frac{c - id}{c - id}\right)$$

where the second bracket is, in fact, unity (i.e. one), and so does not alter the value of the right-hand side of the equation

$$\therefore \frac{z_1}{z_2} = \frac{ac + ibc - iad - i^2 bd}{c^2 + d^2}$$

$$= \frac{(ac+bd)+i(bc-ad)}{c^2+d^2}$$

$$= \left(\frac{ac+bd}{c^2+d^2}\right) + i\left(\frac{bc-ad}{c^2+d^2}\right)$$

$$= v + iw$$

where v, w, in spite of being more elaborate, are real numbers. Hence, the quotient of two complex numbers is a complex number. *Note.* The process for division requires that $z_2 \neq 0$ (i.e. that c and d are not both zero).

Example. Plot the points representing $z_1 = 4+3i$ and $z_2 = 2-5i$ on an Argand diagram. *Measure* the distance between them and *calculate* this distance as a check.

The difference in distance along the real axis is $4-2 = 2$ and the difference along the imaginary axis is $3-(-5) = 8$. We then use Pythagoras's theorem (Fig. 10).

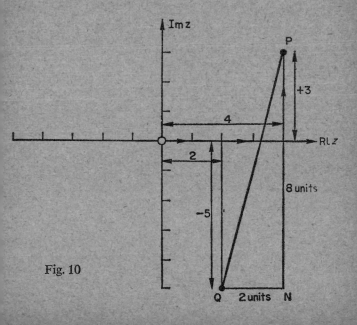

Fig. 10

By measurement $PQ \doteqdot 8 \cdot 2$ units.

By calculation $PQ = \sqrt{QN^2 + NP^2}$

$$= \sqrt{4 + 64} \text{ unit} = \sqrt{68} \text{ unit}$$

$$\doteqdot 8 \cdot 25 \text{ unit.}$$

Exercise 4

1. Which of the following complex numbers are (*a*) real, (*b*) purely imaginary, (*c*) when simplified can only be expressed in the complex form $a + ib$, where a and b are real numbers:

$$i^5, \quad i^6, \quad 4i^7, \quad -3i^4, \quad 4 + 5i^3, \quad 7 - 3i^7, \quad 3i + 8i^3, \quad 2 - i + i^3,$$

$$2 - i - i^3, \quad 2i + \frac{1}{i}.$$

2. Express all the numbers in Qn. 1 above in simplest form.
3. If $z_1 = 1 - i$ and $z_2 = 3 + 2i$, find
$$(a)\ 2z_1 + 5z_2, \quad (b)\ 2z_1 + z_2, \quad (c)\ 4z_1 - 3z_2.$$
4. If $z_1 = 2 + i$ and $z_2 = 3 - 4i$, find
$$(a)\ z_1^2, \quad (b)\ z_1 z_2, \quad (c)\ \frac{z_2}{z_1}, \quad (d)\ \frac{4z_1 + z_2}{3z_1 - 2z_2},$$

expressing the answers in simplest form.
5. Simplify
$$(a)\ \frac{3 + 2i}{3 - i}, \quad (b)\ \left(2i + \frac{1}{i}\right)(1 - i).$$

6. Plot the points representing $6 + 3i$ and $1 - 9i$ on a diagram. *Calculate* the distance between them and *compare* this with the measured distance on the diagram.
7. Show the points $z_1 = 3 + 2i$ and $z_2 = 2 - i$ on a diagram. Demonstrate on the diagram that $z_1 + 2z_2$ is a real number. [*Hint.* $2z_2 = 2(2 - i) = 4 - 2i$, which gives twice the distance from zero on the Rl axis and on the Im axis.]

Although we have only touched on the threshold of complex numbers, the theory of which forms a fairly advanced branch of mathematics, it is desirable to introduce the topic as above to indicate the position of real numbers in the whole realm of numbers. The theory also affords us an interesting introduction to the concept of a vector, developed below and expanded in the next chapter.

4. Cartesian Ordered Pairs

If we consider the real and imaginary parts of a complex number z, where $z = a + ib$, we could write $z = (a, b)$, an *ordered pair* (i.e. one in which the order of the letters a, b is of fundamental importance). We take a to represent a displacement to the right of the starting point—in map work, we would call this an *easting*—and b to represent a displacement above the starting point—a *northing*. This concept exactly accords with the idea (Section 1 above) that multiplication by i causes an anticlockwise rotation of $90°$. The number z itself represents a *translation* from A to C (Fig. 11).

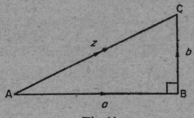

Fig. 11

We have seen (in Section 3 above) that we add two complex numbers z_1 and z_2 as follows:

If $z_1 = a_1 + ib_1$ and $z_2 = a_2 + ib_2$,
then
$$z = z_1 + z_2 = (a_1 + ib_1) + (a_2 + ib_2)$$
$$= (a_1 + a_2) + i(b_1 + b_2)$$

Expressed as ordered pairs this gives

$$(a_1, b_1) + (a_2, b_2) = (a_1 + a_2, b_1 + b_2)$$

If we move Westwards, we are going in the opposite direction to Eastwards. Thus, in Fig. 12, we say $z = (-a, b)$, assuming a, b positive.

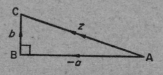

Fig. 12

Similarly, if we go Southwards we are going in the opposite direction to Northwards. Hence, starting from O, the ordered pairs giving $\overrightarrow{OA}$, $\overrightarrow{OB}$, $\overrightarrow{OC}$, $\overrightarrow{OD}$, in Fig. 13, are

$$\overrightarrow{OA} = (a, b),$$
$$\overrightarrow{OB} = (-a, b),$$
$$\overrightarrow{OC} = (-a, -b),$$
$$\overrightarrow{OD} = (a, -b),$$

where a, b are positive.

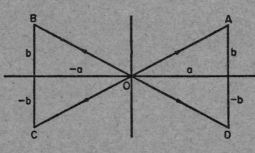

Fig. 13

The notation $\overrightarrow{OA}$ indicates the displacement from O to A in magnitude and direction. This concept leads to the definition of a *vector*, given in Chapter 3 following.

Let us first, however, consider expressing simple algebraic operations with ordered pairs. Let $z_1 = a + ib$, $z_2 = c + id$, then $z_1 = (a, b)$, $z_2 = (c, d)$. We have already seen from above that

$$(a, b) + (c, d) = (a + c, b + d).$$

From section 3 of this chapter, we found

$$(a + ib) - (c + id) = (a - c) + i(b - d),$$
$$(a + ib) \times (c + id) = (ac - bd) + i(bc + ad),$$
$$(a + ib) \div (c + id) = \left(\frac{ac + bd}{c^2 + d^2}\right) + i\left(\frac{bc - ad}{c^2 + d^2}\right).$$

Expressed as ordered pairs, these three additional results yield

$$(a, b) - (c, d) = (a-c, b-d)$$

$$(a, b) \times (c, d) = (ac-bd, bc+ad)$$

$$(a, b) \div (c, d) = \left(\frac{ac+bd}{c^2+d^2}, \frac{bc-ad}{c^2+d^2}\right).$$

Some special cases are worthy of thought.

Example. If $z_1 = (3, -2)$ and $z_2 = (12, 5)$, find $z_1 z_2$ and z_1/z_2.

If $z_1 = (a, b)$, $z_2 = (c, d)$, then $a = 3$, $b = -2$, $c = 12$, $d = 5$. Hence from above, $z_1 z_2 = (ac-bd, bc+ad)$

$$= (36+10, -24+15)$$
$$= (46, -9)$$

and

$$\frac{z_1}{z_2} = \left(\frac{ac+bd}{c^2+d^2}, \frac{bc-ad}{c^2+d^2}\right)$$

$$= \left(\frac{36-10}{144+25}, \frac{-24-15}{144+25}\right)$$

$$= \left(\frac{2}{13}, -\frac{3}{13}\right).$$

Exercise 5

1. Prove that multiplication, or division, of any ordered pair by $(1, 0)$ leaves the ordered pair unaltered. [*Hint.* Let the ordered pair be (a, b).]

2. Find the following, giving the answers as ordered pairs:

(a) $(3, 0) + (2, -5)$,
(b) $(0, 4) - (-2, 4)$,
(c) $(7, 9) - (9, 8)$,
(d) $(0, 1) \times (0, 1)$,
(e) $(a, b) \times (b, -a)$,
(f) $(a, b) \times (b, a)$,
(g) $(x, y) \div (0, 1)$,
(h) $(3, -5) \times (-2, 4)$
(i) $(-3, -2) \div (2, -4)$,
(j) $(a, 0) \div (-2a, 3a)$.

3. Mark on graph paper, using a scale of 1 inch as 1 unit, the points $A(-3, -2)$, $B(2, -1)$, $C(3, 4)$. Draw the path $\vec{AB} + \vec{BC}$ putting in arrows to show the direction of travel. Also draw $\vec{AC}$. Measure $AB + BC - AC$. Compare this result with that obtained by calculation, using Pythagoras's theorem.

$$[\textit{Hint. } AB = (2, -1) - (-3, -2) = (5, 1),$$
$$\therefore AB = \sqrt{5^2 + 1^2} = \sqrt{26}, \text{ etc.}]$$

CHAPTER 3

VECTORS

1. Scalars and Vectors

In our everyday life, although we may give little thought to the matter, we deal with two mathematical entities, namely, quantities which have a definite magnitude but for which direction (say, of movement) has no meaning, and quantities for which direction is of fundamental significance.

It is not difficult to distinguish between these concepts. In physics, for example, length, mass and speed have magnitude but no direction. Such quantities are defined as *scalars*. Other physical units, such as force, momentum and velocity depend very much on the direction in which they act. These are defined as *vectors*.

The reader may be a little puzzled about the distinction between speed and velocity. This is because in ordinary conversation we use mathematical units rather loosely. Strictly speaking, when we say a car has a top *speed* of 93 m.p.h. we are using the unit as a *scalar*, for direction of travel is not being considered. If, however, we say that we travelled due south at 93 m.p.h. (perhaps slightly illegally) then we are using *velocity*, a *vector* (namely: South, 93 m.p.h.). There is a similar distinction between *mass* and *weight*. The former is an inherent property of the molecules forming a body; this does not change no matter where we place the body in space. This is not true of weight, which is the gravitational force attracting the body; the further from the surface of the earth the smaller is this force attracting the body towards the centre of the earth. Furthermore, it is not even the same at different points on the surface of the earth for it acts "vertically downwards". Whatever this means[3]—and it is a convenient phrase—it is not the same at the North Pole as it is, say,

---

[3] There are mathematical complications here—the earth rotates—but they are of no consequence from the point of view of elementary considerations.

at a point on the equator. Even neglecting the fact that the earth is an oblate spheroid and not a true sphere, clearly the vectors are different, for their directions are different (Fig. 14).

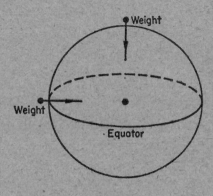

Fig. 14

Many of the applications of vectors lie in the realm of applied mathematics and theoretical physics, but they also have their uses in pure mathematics. We restrict ourselves, in this book, mainly to the concepts of displacement, i.e. change of position from one point to another (introduced at the end of Chapter 2), and of velocity, explained above.

There are two kinds of vectors which we shall use: (i) *generalized vectors*, which are of given magnitude, direction and sense; (ii) *localized vectors*, which have all the properties of generalized vectors but also act through one given point (i.e. along a given line). Sections 2 to 6 will deal mainly with generalized vectors and Section 7 with localized vectors, in this chapter.

2. Coordinates and Vectors

Consider Fig. 15, in which OX and OY are perpendicular axes and P is a point in the plane XOY.

There are two ways of determining the position of P relative to the origin O.

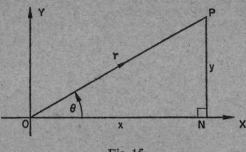

Fig. 15

Firstly, the point P has *rectangular* (cartesian) *coordinates* (x, y), measured from the origin O. This is an ordered pair exactly in the sense of Section 4 of the last chapter, where x represents an easting and y a northing from O to P.

Secondly, we can consider P as being at a distance OP $(= r)$, in the direction θ (theta) measured *anticlockwise* from the direction OX, which for this purpose is called the *initial line*. We say that P has polar coordinates (r, θ). A word of warning is necessary here. *We cannot manipulate two or more pairs of polar coordinates as ordered pairs in the same way as we can rectangular coordinates.*

Further, we can think of P as being an end point of the vector $\overrightarrow{OP}$, which has magnitude r, direction θ (polar coordinates), and components x, y (rectangular coordinates). The significance of the arrow-head above the letters in the vector symbol is that we are considering a translation from O to P, i.e. $O \rightarrow P$. Thus, $\overrightarrow{PO}$ would be a vector of the same magnitude, length OP, but from P to O, i.e. $P \rightarrow O$. This is discussed more fully on p. 44. We avoid the use of the arrow-head in the reverse direction ($\leftarrow$) as this tends to cause confusion.

Elementary trigonometry[4] immediately shows us that there are four fundamental relationships among the four variables x, y, r, θ. Three of these are trigonometric and one is algebraic.

We have, in Fig. 16,

---

[4] The reader who has a little difficulty with simple trigonometrical ratios is referred to *Teach Yourself Trigonometry*, E.U.P., or some similar elementary work. Only the simplest ideas are necessary.

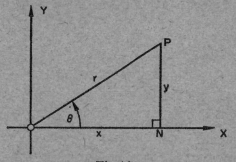

Fig. 16

$$\frac{ON}{OP} = \cos\theta \quad \text{and} \quad \frac{NP}{OP} = \sin\theta,$$

i.e. $$\frac{x}{r} = \cos\theta \quad \text{and} \quad \frac{y}{r} = \sin\theta.$$

These give the relationships

$$x = r\cos\theta \dots (1), \quad y = r\sin\theta \dots (2),$$

which can be used to convert from rectangular to polar coordinates.
Also, in the same figure,

$$OP^2 = ON^2 + NP^2 \quad \text{and} \quad \tan\theta = \frac{NP}{ON},$$

i.e. $$r^2 = x^2 + y^2 \dots (3), \quad \tan\theta = \frac{y}{x} \dots (4),$$

which serve to convert from polar to rectangular coordinates.

The four relationships are not all independent. Given any two, the others may be derived from them.

For instance, from (1) and (2), on squaring and adding,

$$x^2 + y^2 = r^2\cos^2\theta + r^2\sin^2\theta$$
$$= r^2(\cos^2\theta + \sin^2\theta)$$
$$= r^2, \quad \text{which is relationship (3).}$$

Also, by dividing (2) by (1), we have

$$\frac{y}{x} = \frac{r\sin\theta}{r\cos\theta} = \tan\theta, \quad \text{which is relationship (4).}$$

For work in this book, r will *usually* be taken as positive and θ to be an angle lying between $0°$ and $360°$, measured anticlockwise from OX. The values of x and y may be taken as positive or negative. There is no difficulty about taking r as positive.

If, for example, a point P has rectangular coordinates $(a, -b)$, where a and b are positive, then

$$r^2 = a^2+(-b)^2 = a^2+b^2 \Rightarrow r = \pm\sqrt{(a^2+b^2)},$$

and by convention we take $r = +\sqrt{(a^2+b^2)}$.

Fig. 17 illustrates the four possible cases.

2nd quadrant   1st quadrant

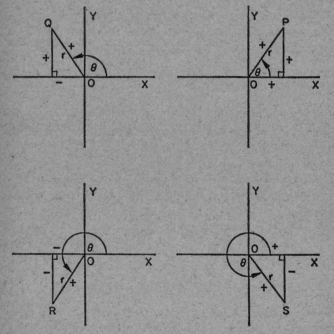

3rd quadrant   4th quadrant

Fig. 17

These cases accord exactly with elementary trigonometry.

Example. If $\vec{OP} = (12, 5)$ and $\vec{OQ} = (12, -5)$, find the polar coordinates of P and Q.

A diagram, which need not be to scale, is helpful here (Fig. 18).

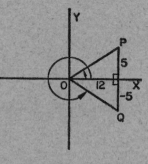

Fig. 18

For OP, we have $x = 12$, $y = 5$,

$$\therefore r^2 = x^2 + y^2 = 144 + 25 = 169 \Rightarrow r = 13$$

and

$$\tan \theta = \frac{y}{x} = \frac{5}{12} \approx 0.417 \Rightarrow \theta \approx 23°.$$

Hence the polar coordinates of P are $(13, 23°)$. Alternatively, we can say, $\vec{OP} = (13, 23°)$. Electrical engineers often write this as $(13, \angle 23°)$.

For $\vec{OQ}$, we have $x = 12$, $y = -5$,

$$\therefore r^2 = x^2 + y^2 = 12^2 + (-5)^2 = 144 + 25 = 169 \Rightarrow r = 13,$$

as before, but this time

$$\tan \theta = -\frac{5}{12} \approx -0.417 \Rightarrow \theta \approx 360° - 23° = 337°.$$

(The diagram clearly indicates this, but without it we might have been in trouble, for from Fig. 17 the tangent is also negative in the 2nd quadrant as well as in the 4th, and we could have obtained the incorrect result $180° - 23° = 157°$. There is an alternative mathematical method of discrimination by considering the signs of sine, cosine and tangent, but the reader will find that a diagram is simpler).

Hence the polar coordinates of Q are $(13, 337°)$. Alternatively $\vec{OQ} = (13, 337°)$.

Example. If $\vec{OR} = (4, 30°)$, find the rectangular coordinates of R.

We have $\quad x = r \cos \theta = 4 \cos 30° = 4 \times \dfrac{\sqrt{3}}{2} = 2\sqrt{3}$

and $\qquad\qquad y = r \sin \theta = 4 \sin 30° = 4 \times \dfrac{1}{2} = 2.$

$\therefore$ the required coordinates of R are $(2\sqrt{3}, 2)$. Alternatively, $\vec{OR} = (2\sqrt{3}, 2)$.

Aide Memoire. The following table should help the reader's memory. The results are derived from any standard textbook of elementary trigonometry.

Ratio \ Angle	0°	30°	45°	60°	90°
sin	0	$\dfrac{1}{2}$	$\dfrac{1}{\sqrt{2}}$	$\dfrac{\sqrt{3}}{2}$	1
cos	1	$\dfrac{\sqrt{3}}{2}$	$\dfrac{1}{\sqrt{2}}$	$\dfrac{1}{2}$	0
tan	0	$\dfrac{1}{\sqrt{3}}$	1	$\sqrt{3}$	∞

Exercise 6

1. Express, in rectangular coordinates, the vectors given by the following polar coordinates: (*a*) $(10, 60°)$, (*b*) $(8, 45°)$, (*c*) $(3, 0°)$, (*d*) $(5, 90°)$, (*e*) $(1, 30°)$, (*f*) $(2, 120°)$, (*g*) $(6, 225°)$.

2. Express, in polar form, the vectors given by the following cartesian ordered pairs: (*a*) $(3, 4)$, (*b*) $(3, -4)$, (*c*) $(-3, 4)$, (*d*) $(-3, -4)$.

3. P is the point $(5, 2)$ and Q is $(3, 4)$. Express $\vec{PQ}$ (*a*) as a cartesian ordered pair, (*b*) in polar form.

4. $\vec{OA} = (1, -3)$, $\vec{OB} = (2, 1)$, $\vec{OC} = (-1, 2)$. (i) Find the excess by which the length $(AB + BC)$ exceeds AC. (ii) Express $\vec{AB}$ and $\vec{AC}$ in polar form. (iii) Find angle BAC. [*Note.* A sketch is helpful but is not needed for actual calculations.]

5. $OH = (3, 90°)$, $\vec{OK} = (4, 150°)$. Draw a diagram to illustrate $\vec{HK}$. Calculate the length of HK. If $\vec{ON}$ is perpendicular to $\vec{HK}$, find the bearing of $\vec{ON}$ from the initial line. [*Hint.* Convert to cartesian ordered pairs.]

3. The Addition of Vectors

When, in arithmetic, we have added two scalars together, we have always expected to have a simple arithmetical total. Thus

$$3 \text{ m} + 2 \text{ m} = 5 \text{ m}$$
and
$$£4·60 + £2·55 = £7·15.$$

Such a result does not necessarily apply when vectors are added, for here we have to take direction as well as magnitude into account.

Consider Fig. 19, in which $\vec{AB} = (a_1, a_2)$, $\vec{AC} = (b_1, b_2)$. We complete the parallelogram $ABDC$, then AC is equal and parallel to BD.

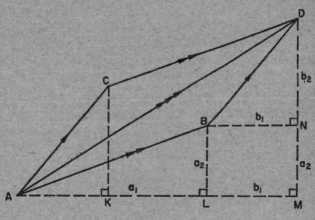

Fig. 19

We have $\vec{AB} + \vec{AC} = \vec{AB} + \vec{BD}$
$$= (a_1, a_2) + (b_1, b_2)$$
$$= (a_1 + b_1, a_2 + b_2) \qquad \text{(from page 33)}$$
$$= (AM, MD)$$
$$= \vec{AD}.$$

Clearly the length AD is less than the length $(AB+BD)$. This follows at once from the Euclidean theorem that the sum of the lengths of any two sides of a triangle is greater than the length of the third side. Thus the rule for the addition of two vectors is not the same as for the addition of scalars, in magnitude. Neither is it in direction, for we see that by adding $\overrightarrow{AB}$ and $\overrightarrow{BD}$ we have arrived at the vector $\overrightarrow{AD}$ formed by joining the open end-points of $\overrightarrow{AB}$ and $\overrightarrow{BD}$ in the same order of travel.[5]

It will be observed that, although we have carefully indicated the movement from the starting point of a vector to its finishing point, we have not restricted ourselves as to the location of the starting point. Thus, parallel vectors of equal length are equal, e.g. in Fig. 20, $\overrightarrow{AB} = \overrightarrow{PQ}$.

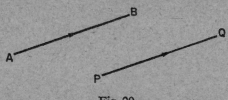

Fig. 20

Reversing a vector changes its sign but does not affect its magnitude (length). This is a reasonable state of affairs for, in Fig. 21, if we take $\overrightarrow{RS}+\overrightarrow{SR}$, we proceed from R to S and back again to R. Thus we have not got anywhere.

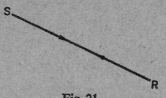

Fig. 21

[5] There is, of course, the special case when $\overrightarrow{AB}$ and $\overrightarrow{BD}$ are in the same sense along the same straight line. The magnitude AD is then $AB+BD$ and the direction of $\overrightarrow{AD}$ is that of $\overrightarrow{AB}$ and $\overrightarrow{BD}$.

The legwork may be considerable but the vector change is zero. We say, therefore, that

$$\vec{RS} + \vec{SR} = 0$$

i.e. $$\vec{SR} = -\vec{RS}.$$

If $\vec{RS} = (x, y)$, then $\vec{SR} = (-x, -y)$.

This is easily verified, for

$$\vec{RS} + \vec{SR} = (x, y) + (-x, -y) = (0, 0)$$
$$= 0.$$

In Fig. 20, we say that $\vec{AB}$ and $\vec{PQ}$ are in the same sense, but in Fig. 21, $\vec{RS}$ and $\vec{SR}$ are in opposite sense, i.e. are measured in opposite ways along the same line (or parallel lines).

In polar coordinates, if $RS = (r, \theta)$, then $\vec{SR} = (r, 180° + \theta)$, i.e. the same displacement in the reverse direction. The reader is, however, again warned not to attempt to add ordered pairs of polar coordinates in the way he has added rectangular coordinates. We *can* get round the difficulty by considering $\vec{SR} = (-r, \theta)$ and the section 4 below indicates how this would help.

4. The Multiplication of a Vector by a Scalar

The multiplication of a vector by a scalar does not alter the direction of the former but does change its magnitude. Suppose $\vec{AB} = (x, y)$, then, say, $3\vec{AB} = 3(x, y) = (3x, 3y).$

or, in polars, $3\vec{AB} = (3r, \theta).$

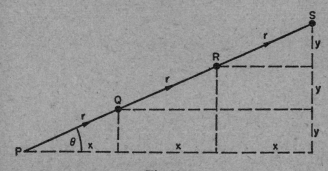

Fig. 22

Fig. 22 makes this clear, for $\vec{PQ} = \vec{QR} = \vec{RS}(= \vec{AB})$

$$\therefore\ 3\vec{AB} = (x, y) + (x, y) + (x, y)$$

$$= (3x, 3y)\quad \text{(by addition of rectangular coordinates).}$$

Furthermore, clearly the direction of travel has not changed, but the total distance $PS = PQ + QR + RS = 3r$.

$$\therefore\ 3\vec{AB} = (3r, \theta).$$

If again $\vec{AB} = (x, y)$, let us now multiply by -1. In rectangular coordinates, $-\vec{AB} = (-x, -y)$, as we have already seen. In polar coordinates, $-\vec{AB} = (-r, \theta)$, on multiplying the r component by -1 and on leaving θ alone. This suggests that we can add polar coordinates if, and only if, the angle θ is the same in each ordered pair we are considering. This is logical, for in this case the vectors are along the same or parallel lines and the conclusion accords with footnote 5 on page 44.

5. Abbreviated Notation

Instead of representing a vector by its end points, with an arrow above indicating the *sense* of the vector, we can use a single small letter (lower case) in Clarendon type. Thus, in Fig. 23, if $\vec{AB} = \mathbf{a}$, then $\vec{BA} = -\mathbf{a}$.

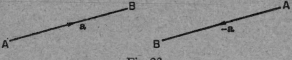

Fig. 23

We shall make use of this notation in the following exercises.

Example. In Fig. 24, find $\mathbf{x}$ in terms of $\mathbf{a}$ and $\mathbf{b}$.

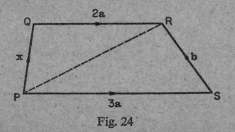

Fig. 24

Draw the diagonal PR,

then
$$\vec{PR} = \vec{PQ} + \vec{QR}$$
$$= x + 2a$$

Also
$$\vec{PR} = \vec{PS} + \vec{SR}$$
$$= 3a + b$$

$\therefore \qquad\qquad x + 2a = 3a + b$

i.e. $\qquad\qquad x = a + b.$

It is of interest to note in this example that $PQRS$ is a trapezium, for $\vec{QR} = 2a$ and $\vec{PS} = 3a$, both scalar multiples of a, the same vector, and therefore parallel to one another.

This example can be solved in a slightly different way, one which brings out a fundamental principle in adding vectors. We must be particularly careful to add only vectors which are end to end and then only in order proceeding from the starting point.[6]

Thus, in Fig. 24, we have

$$\vec{PQ} + \vec{QR} + \vec{RS} + \vec{SP} = 0,$$

for we have, in adding these vectors, returned to our starting point.

Now $\vec{PQ} = x$, $\vec{QR} = 2a$, $\vec{RS} = -\vec{SR} = -b$, $\vec{SP} = -\vec{PS} = -3a$, therefore $\qquad x + 2a - b - 3a = 0,$

giving $\qquad\qquad x = a + b,$ as before.

It is worth emphasising this point by examining the Figs. 25 (a) and 25(b).

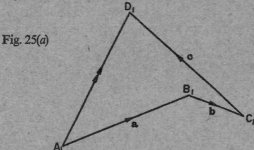

Fig. 25(a)

[6] The mathematician gets round this little problem adroitly but the fledgling is advised to adhere to this principle until he is ready to leave the nest.

In Fig. 25(a), we have

$$\overrightarrow{A_1 D_1} = \overrightarrow{A_1 B_1} + \overrightarrow{B_1 C_1} + \overrightarrow{C_1 D_1},$$
$$= \mathbf{a} + \mathbf{b} + \mathbf{c}$$

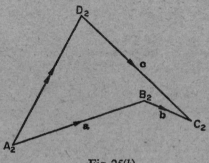

Fig. 25(b)

In Fig. 25(b), we have

$$\overrightarrow{A_2 D_2} = \overrightarrow{A_2 B_2} + \overrightarrow{B_2 C_2} + \overrightarrow{C_2 D_2}$$
$$= \mathbf{a} + \mathbf{b} - \mathbf{c}.$$

The double arrow is used to indicate the *resultant* of the three vectors. It can be used, of course, to represent the resultant of any number of vectors added in this way. The double arrow notation is not, however, universally used.

Exercise 7

1. If $\mathbf{a} = (3, 2)$, $\mathbf{b} = (-4, 3)$, $\mathbf{c} = (2, -1)$,
   $\mathbf{d} = (-1, 0)$, $\mathbf{e} = (0, -5)$, $\mathbf{f} = (-3, -2)$,
   find   (i) $\mathbf{a} + \mathbf{b} + \mathbf{d}$,   (ii) $\mathbf{c} - \mathbf{f}$,   (iii) $2\mathbf{a} + \mathbf{c} + 2\mathbf{b}$,
   (iv) $\mathbf{a} + \mathbf{d} + \mathbf{f}$,   (v) $3\mathbf{a} + \mathbf{b} - \mathbf{e} + \mathbf{f}$,   (vi) $2\mathbf{a} + \mathbf{b} + \mathbf{c} + 4\mathbf{d}$.

2. If $\mathbf{a} = (2, 1)$, $\mathbf{b} = (1, -3)$, $\mathbf{c} = (-1, 2)$, draw separate diagrams on graph paper to indicate
   (i) $\mathbf{a} + \mathbf{b}$,   (ii) $\mathbf{a} - \mathbf{b}$,   (iii) $\mathbf{a} + \mathbf{b} - \mathbf{c}$,
   (iv) $\mathbf{b} - \mathbf{c}$,   (v) $\mathbf{b} + 2\mathbf{c}$,   (vi) $\mathbf{a} - 2\mathbf{c}$.

3. If $\overrightarrow{OP} = (3, -2)$ and $\overrightarrow{PQ} = (1, 5)$, find the bearing and distance of Q from O. Write the result in the form of polar coordinates.

4. An aircraft, flying in still air, is searching for wreckage and traverses the path *OPQRST*, where $\overrightarrow{OP} = (1, 2)$, $\overrightarrow{PQ} = (2, -1)$, $\overrightarrow{QR} = (-1, -5)$, $\overrightarrow{RS} = (-3, -1)$, $\overrightarrow{ST} = (3, 3)$. Find how far it will be to return from *T* to base *O*. In what direction will the aircraft steer? Draw, on graph paper, the path traced out, indicating with arrows the sections *OP*, *PQ*, etc. traversed.

5. In the $\triangle ABC$, if $\overrightarrow{AB} = \mathbf{p}$ and $\overrightarrow{BC} = \mathbf{q}$, what is $\overrightarrow{AC}$? [*Hint*. Draw a sketch.]

6. If, in $\triangle XYZ$, $\overrightarrow{XY} = \mathbf{a}$ and $\overrightarrow{XZ} = \mathbf{b}$, what is $\overrightarrow{YZ}$? [*Hint*. One vector is in the opposite direction to that needed for addition. We therefore change its sign before adding.]

7. In parallelogram *ABCD*, $\overrightarrow{AB} = \mathbf{a}$ and $\overrightarrow{BC} = \mathbf{b}$. Write down the diagonal vectors $\overrightarrow{AC}$ and $\overrightarrow{BD}$ in terms of $\mathbf{a}$ and $\mathbf{b}$. What is $\overrightarrow{DB}$? [The results have useful applications in vector analysis.]

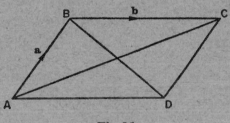

Fig. 26

8. In Fig. 27, find **a** in terms of the other vectors.

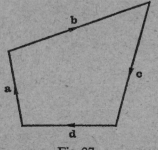

Fig. 27

9. In Fig. 28, find $\vec{AC}$, $\vec{AD}$ and **e** in terms of the other vectors **a**, **b**, **c**, **d**.

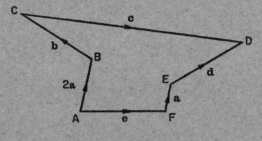

Fig. 28

10. In Fig. 29, calculate $\vec{OM}$ in terms of **a** and **b** only.
[*Note.* $\vec{AM} = \vec{MB} = \mathbf{c}$.]

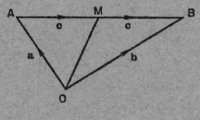

Fig. 29

6. Applications of Vectors to Problems on Velocity

Suppose that Davy Jones, who has had a heavy night, is loafing in a deck-chair on board a ship travelling due East at 12 knots. Suddenly Davy thinks that he sees a great auk fly past. (What an imagination! The bird has been extinct for years.) He leaves his chair and walks to the stern at 3 knots to examine his "auk" more closely. His true velocity over the sea, whilst he is walking, is $(12-3)$ knots, i.e. 9 knots, due East (Fig. 30).

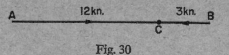

Fig. 30

$$\vec{AB}+\vec{BC} = (12, \text{East})+(-3, \text{East})$$
$$= (9, \text{East}).$$

We can add the polar ordered pairs as the *course*, due East, is the same in both vectors. It will be observed that the vectors $\vec{AB}$, etc., in this case represent not distance but velocity.

Suppose now that Davy, fully recovered from his hangover, turns about and walks towards the bows of the ship at 3 knots. His new velocity over the sea (Fig..31) is given by

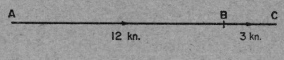

Fig. 31

$$\vec{AB}+\vec{BC} = (12, \text{East})+(3, \text{East})$$
$$= (15, \text{East}).$$

He is thus travelling due East at 15 knots, i.e. (15 kn., 090°).

Tired of his perambulations from one end of the ship to the other, Davy now walks from starboard to port at his speed of 3 knots. The problem of his resultant velocity over the sea is now more complex, for he has one velocity, of 12 knots due East, and another, of 3 knots due North.

It is more easily seen if we imagine Davy to be represented by a small object placed on a piece of transparent paper on top of this page. The transparent paper is moved 4 units to the right and at the same time the object is moved 1 unit to the top of the paper (*Note*. 12 knots: 3 knots = 4: 1, so we are preserving Davy's problem). The experiment is repeated, the paper now having moved a total of 8 units to the right and the object 2 units to the top, and so on. Clearly, in Fig. 32, the *actual path* of the object across the page is along the line AC_1C_2 ...

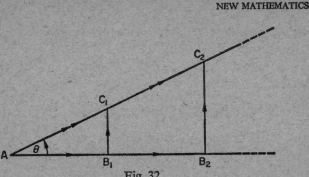

Fig. 32

In vector analysis, the theory of vectors, it is usual to represent a *unit vector along* the axis OX by the Clarendon type **i** and a *unit vector along* the perpendicular axis OY by **j**. If v is Davy's resultant velocity when walking from starboard to port across his ship, which is travelling due East, we have (Fig. 33)

$$\mathbf{v} = 12\mathbf{i} + 3\mathbf{j}.$$

Hence
$$v = \sqrt{12^2 + 3^2} = \sqrt{153}$$
$$\simeq 12{\cdot}4 \text{ (knots)}.$$

and
$$\tan \theta = \frac{3}{12} = 0{\cdot}25$$

giving
$$\theta \simeq 14°, \text{ from tables.}$$

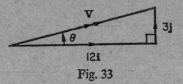

Fig. 33

Thus Davy's true velocity is   N76°E, 12·4 knots.

The reader who is familiar with navigation may have observed that there is a slight difficulty of notation here. In polar coordinates, direction is measured *anticlockwise* from OX and may vary from 0° to 360° (more, if several rotations are considered), whereas in navigation direction (course *or* bearing) is measured *clockwise* from due North and may vary from 0° to 360°. In consequence, we have adopted the older representation of *course* or *bearing* here, i.e. from North to East or West and from South to East or West, e.g. N15°W, S72°W, S33°E.

Example. A river is flowing at 4 km/h. Brown, who ought to know better, steers a motor-boat directly across the river, which is 110 m wide, at 8 km/h. How far downstream does he land in the opposite bank?

On the return journey, Brown decides to take his motorboat directly across the river. What course must he steer at the same speed? What is the speed made good?

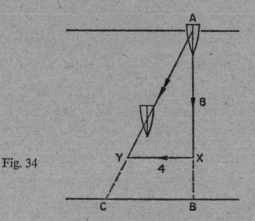

Fig. 34

(a) *Outward Journey.*

In Fig. 34, AX is the velocity of the motor-boat on the water, and XY is the velocity of the river. Then AY is the *actual* velocity of the motor-boat (its velocity made good).

By similar triangles CBA, YXA

$$\frac{CB}{AB} = \frac{YX}{AX}$$

$$\therefore \frac{CB}{110} = \frac{4}{8} \Rightarrow CB = 55.$$

Hence the distance downstream is 55 m.

It is interesting to observe that we have compared a triangle of displacements with a triangle of velocities.

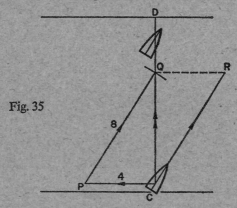

Fig. 35

(b) *Return Journey.*
This is more subtle. During the time the river moves the motor-boat 4 units of distance (whatever they may be) downstream, the boat itself travels 8 units (of the same type). Draw $CP = 4$ units (Fig. 35) downstream from C, the starting point of the return journey. Centre P, radius 8 units, draw an arc PQ to cut CD, the resultant path directly across the river (which is what is needed in the question). $\vec{PQ}$ is the vector for the motor-boat's velocity. The actual course steered by the boat is $\vec{CR} = \vec{PQ}$, starting at C, but the reader will recall that equal vectors only need to be equal in size and parallel in direction and sense.

We have $\cos\theta = \dfrac{PC}{PQ} = \dfrac{4}{8} = 0.5 \Rightarrow \theta = 60°,$

so the course to steer is in a direction 60° with the river bank viewed upstream.

The speed made good is directly across the river and is CQ. Using Pythagoras's theorem, we have

$$CQ^2 = PQ^2 - PC^2$$
$$\Rightarrow CQ = \sqrt{(PQ^2 - PC^2)} = \sqrt{(64 - 16)}$$
$$= \sqrt{48} = 4\sqrt{3} \simeq 6.93$$

∴ speed made good is 6·93 km/h.

Exercise 8

1. An aircraft flies at a constant air speed (i.e. speed through the air) of 120 knots, from West to East and back again, the distance each way being 210 nautical miles. If there is an East wind (i.e. a wind *from* the East) of 30 knots, find (*a*) the ground speed on the outward journey, (*b*) the ground speed on the return journey, (*c*) the total time of flight of the aircraft.

2. An aircraft flies in a direction due South at an air speed of 108 knots. The wind is blowing from the West at 45 knots. What are the aircraft's ground speed and course made good?

3. The aircraft in Qn. 2 above again travels at an air speed of 108 knots and the wind still blows from the West at 45 knots. What course must be steered for the aircraft to proceed due South over the ground?

4. Two ships leave harbour simultaneously. One proceeds due North at 12 knots and the other due West at 8 knots. What is the constant bearing of the first ship from the second? How far are they apart after 2 hours?

5. Abbotsville is 200 km North East of Bishopsbury. The wind is blowing from the North at 40 km/h. An aircraft leaves Abbotsville and flies, at a constant airspeed of 160 km/h, to Bishopsbury and later returns to base. Draw to scale on graph paper a diagram showing the courses to steer on the outward and homeward journeys respectively. Measure these results and also measure the ground speeds for the two journeys. How long will the journey take each way?

7. Localised Vectors

Hitherto we have been concerned with processes which apply to vectors in general. We now consider problems in which the point from which we measure is of paramount importance and we shall in fact measure all our vectors (in this section) from an origin, even though we may not assign the letter O to it in some instances.

Suppose we know that the point P on a line $\vec{PQ}$ is given by **a** relative to origin O, i.e. $\vec{OP} = \mathbf{a}$. If $\vec{PQ} = \mathbf{b}$, any other vector length PR along PQ is $k\mathbf{b}$ where k is a scalar multiplier of **b**. The localised vector giving the position of R relative to O is given by Fig. 36.

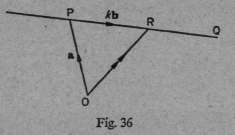

Fig. 36

$$\vec{OR} = \vec{OP} + \vec{PR}$$
$$= \mathbf{a} + k\mathbf{b}.$$

If k is positive, R is on the same side of P as Q. If k is negative, R is on the opposite side of P from Q.

Example. $ABCD$ is a square of side one unit. P is the midpoint of AB and Q is the point of trisection of BC nearer to B. If AQ and DP intersect at R, find AR.

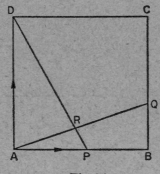

Fig. 37

Let[7] $\vec{AB} = \mathbf{i}$ and $\vec{AD} = \vec{BC} = \mathbf{j}$, then $\vec{AP} = \frac{1}{2}\mathbf{i}$ and $\vec{BQ} = \frac{1}{3}\mathbf{j}$. Take A as the origin, hence
$$\vec{AQ} = \mathbf{i} + \tfrac{1}{3}\mathbf{j}$$

[7] It will be remembered that $\mathbf{i}$ and $\mathbf{j}$ are defined on p. 52.

∴ any point R on $\vec{AQ}$ is given by the localised vector, measured from A,

$$\vec{AR} = k(\mathbf{i}+\tfrac{1}{3}\mathbf{j}) \tag{1}$$

Now $\vec{DP} = \vec{DA}+\vec{AP} = \vec{AP}-\vec{AD} = \tfrac{1}{2}\mathbf{i}-\mathbf{j}$

∴ any point R on $\vec{DP}$ is $l(\tfrac{1}{2}\mathbf{i}-\mathbf{j})$, measured from D; so measured from A, we have

$$\vec{AR} = \vec{AD}+\vec{DR}$$
$$= \mathbf{j}+l(\tfrac{1}{2}\mathbf{i}-\mathbf{j}) \tag{2}$$

If R is the same point, we have from (1) and (2)

$$k\mathbf{i}+\tfrac{1}{3}k\mathbf{j} = \mathbf{j}+\tfrac{1}{2}l\mathbf{i}-l\mathbf{j},$$

where k and l are scalars to be determined.

Hence $\qquad (k-\tfrac{1}{2}l)\mathbf{i}+(\tfrac{1}{3}k-1+l)\mathbf{j} = 0.$

Now this is only possible if the coefficients of $\mathbf{i}$ and $\mathbf{j}$ are *both* zero, for $\mathbf{i}$ and $\mathbf{j}$ are vectors in different directions. (They are, in fact, perpendicular but it would not matter if they were not, provided that they were not parallel).

∴ $\qquad k-\tfrac{1}{2}l = 0 \quad$ and $\quad \tfrac{1}{3}k-1+l = 0.$

Hence $\qquad l = 2k \quad$ and $\quad k+3l = 3.$

Substituting for l in terms of k,

$$k+6k = 3 \Rightarrow 7k = 3 \Rightarrow k = \tfrac{3}{7}.$$

Substituting for k in equation (1) above

$$\vec{AR} = \tfrac{3}{7}\mathbf{i}+\tfrac{1}{7}\mathbf{j}.$$

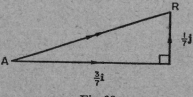

Fig. 38

Hence, by Pythagoras's theorem, remembering **i** and **j** are perpendicular vectors (and this time it *is* helpful!) (Fig. 38)

$$AR^2 = \left(\frac{3}{7}\right)^2 + \left(\frac{1}{7}\right)^2 = \frac{9+1}{49} = \frac{10}{49}$$

$$\therefore AR = \frac{\sqrt{10}}{7} \simeq \frac{3 \cdot 162}{7} \simeq 0 \cdot 45 \text{ unit.}$$

Once the method is understood many of the lines of working in this example can be omitted.

Exercise 9

Most, but not all, of the questions are concerned with localised vectors.

1. O is the origin, $\vec{OA} = \mathbf{a}$ and $\vec{AP} = \mathbf{r}$. The point Q lies on $\vec{AP}$. Express $\vec{OQ}$ in terms of $\mathbf{a}$, $\mathbf{r}$ and a scalar number k. [*Hint.* A sketch is helpful.]

2. In Fig. 39, $\vec{OA} = \mathbf{a}$, $\vec{OB} = \mathbf{b}$. Write $\vec{AB}$ in terms of $\mathbf{a}$ and $\mathbf{b}$.

   P is any point on $\vec{AB}$. If $AP:AB = k:1$, find $\vec{OP}$ in terms of $\mathbf{a}, \mathbf{b}, k$.

[*Hint.* As $AP:AB = k:1$; $AP = k.AB \Rightarrow \vec{AP} = k.\vec{AB}$.]

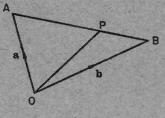

Fig. 39

3. Prove, by vectors, that the line joining the middle points of two sides of a triangle is parallel to the third side and equal in length to one-half of it.

[*Hint.* If the triangle is ABC, let $\vec{AB} = \mathbf{a}$, $\vec{AC} = \mathbf{b}$.]

4. Prove, by vectors, that the diagonals of a parallelogram bisect each other.

5. In the parallelogram $ABCD$, E is the midpoint of AB and F is the midpoint of BC. The lines AF and DE intersect at K. Prove that $AK:KF = 2:3$.

6. ABC is a triangle. The points L, M, N are the midpoints of BC, CA, AB respectively. Prove that AL, BM, CN are concurrent (i.e. meet at one point). If this point is G, prove that $AG:GL= 2:1$. [*Hint.* Take A as origin. Let $\overrightarrow{AB} = \mathbf{c}$, $\overrightarrow{AC} = \mathbf{b}$, say.]

CHAPTER 4

SCALES OF NOTATION

1. Number Systems

In ordinary arithmetic we use the denary (decimal) system (or scale of 10), although some people have wondered why, with a coinage which hitherto has consisted basically of 12d. = 1s., in Britain, we did not adopt the duodecimal system (i.e. scale of 12) in the past. There is now, however, a rapid movement throughout the civilised world towards universal adoption of the denary system. Whether a coinage reads

$$100p. = £1, \quad 100c. = \$1, \quad 100pf. = 1 M.,$$

is immaterial. International financial transactions will be greatly simplified by this general practice. So also will be the arithmetical studies of our youthful slaves in school.

The denary system, in spite of its obvious advantages in general life, is not always ideally suited to mathematical needs. We shall see that the binary (scale of 2) and octal (scale of 8) systems are of considerable importance. Just as the revival of interest in Boolean algebra (Chapter 6) has arisen because of the invention of electronic computers, so likewise has the new enthusiasm for binary arithmetic been dependent on the fact that these machines require the presentation in the scale of 2 of material to be analysed. This is because the myriad complex circuits of a computer work on a YES–NO system. Either a circuit accepts information (YES: equivalent to 1) or does not (NO: equivalent to 0), and so only the digits 1 and 0 are needed. For computing, therefore, numbers are converted to the binary scale.

We start, however, by continuing our studies of the seximal scale, already commenced in Chapter 1, Section 1, not only because it was historically earlier but also because in some ways it is easier to follow. The numbers in this scale, we recall (p. 14), are 1, 2, 3, 4, 5, 10, 11, 12, 13, 14, 15, 20, 21, ...

2. The Seximal Scale

In order to convert from the denary scale to the seximal scale, we need to reflect on what a number such as, say, 5473 means. In the denary scale the number is

Thousands	Hundreds	Tens	Units
5	4	7	3

i.e. $5473 = 5.10^3 + 4.10^2 + 7.10 + 3$
$= 3 + 7.10 + 4.10^2 + 5.10^3$ (on reversing the order).

(N.B. The reader is reminded that, say, 7.5 means 7×5, i.e. 35, and it is *not* the same as 7·5, which means 7 point 5, i.e. $7\frac{1}{2}$.)

In general, a number N can be written, in the denary scale, as

$$N = a_0 + a_1.10 + a_2.10^2 + a_3.10^3 + \ldots$$

where every a is a member of the set[8] $\{0, 1, 2, 3 \ldots 9\}$, i.e. $a_r \in \{0, 1, 2 \ldots 9\}$ for $r = 0, 1, 2 \ldots$ The digits $a_0, a_1, a_2 \ldots$ are called place-holders for the units, tens, hundreds, thousands etc. and we continue to write them down in a number until we have as many as we need. If we go too far, of course, the a's become zeros and we merely waste our time!

There is no reason why the number should be expressed as a series of powers of ten any more than as a series of powers of some other number. In the seximal scale, in fact, we have

$$N = b_0 + b_1.6 + b_2.6^2 + b_3.6^3 + \ldots$$

where every $b_r \in \{0, 1, 2, 3, 4, 5\}$ for all $r = 0, 1, 2 \ldots$

Throughout this chapter denary numbers will be without suffixes but *numbers in other scales will have suffixes* (e.g. 47_6 would indicate "47 [*read* four-seven, *not* forty-seven] in the scale of 6") or will be clearly indicated in some other way.

Let us now convert the number 5473 to the scale of 6

$$
\begin{array}{r}
6)5473 \\
\hline
6)\ \ 912 + 1 \\
\hline
6)\ \ 152 + 0 \\
\hline
6)\ \ \ 25 + 2 \\
\hline
\ \ \ \ 4 + 1 \\
\end{array}
$$

[8] Set notation is explained in Chapter 6. The symbol $\in$ means "is a member of".

We read the answer round the edge as indicated

$$5473 = 41201_6$$

This is easily checked, for

$$41201_6 = 4.6^4 + 1.6^3 + 2.6^2 + 0.6 + 1$$
$$= 5184 + 216 + 72 + 0 + 1$$
$$= 5473.$$

The verification shows how simple it is to convert from a non-decimal scale to the decimal scale.

The seximal multiplication table is interesting as an example

×	1	2	3	4	5
1	1	2	3	4	5
2	2	4	10	12	14
3	3	10	13	20	23
4	4	12	20	24	32
5	5	14	23	32	41

Example. Multiply $(25 \times 34)_6$, the notation meaning that the numbers are in the scale of 6.

Using the table

$$
\begin{array}{r}
34 \\
\times 25 \\
\hline
302 \\
112 \\
\hline
1422 \\
\end{array}
\text{ (scale of 6)}
$$

Check: $1.6^3 + 4.6^2 + 2.6 + 2 = 216 + 144 + 12 + 2 = 374.$

$$25_6 = 17; \quad 34_6 = 22; \quad 17 \times 22 = 374.$$

The check by conversion to the denary system is not necessary once the basic processes have been mastered in other systems.

Exercise 10

The numbers in questions 1–3 are all in the seximal scale.
1. Find

        (a) $31 + 52$, (b) $254 - 125$, (c) $315 - 424 + 105$.

2. Multiply

        (a) 41 by 32, (b) 405 by 43, (c) 1543 by 305.

3. Divide

        (a) 433 by 3, (b) 1103 by 25, (c) 10045 by 41.

4. Convert the following to the denary scale:

        (a) 3154_6, (b) 20403_6, (c) $100,001_6$.

5. Convert the following denary numbers to the scale of six:

        (a) 1479, (b) 10101, (c) 73942.

6. Check the results of questions 1–3 above, by converting the given numbers and the results to the scale of 10.

7. How many digits are needed to express (a) the denary number 10^7 in the seximal scale, (b) the seximal number 14^{11} in the denary scale?

3. The Binary Scale

The relevance of the binary scale to computing has already been mentioned. It is a very easy system to use, for the only digits are 0 and 1. Counting proceeds thus

Scale of 10	1	2	3	4	5	6	7	8 ...
Scale of 2	1	10	11	100	101	110	111	1000...

For example, 6 (in the denary scale) is given by

$$6 = 1.2^2 + 1.2 + 0.1$$
$$= 110 \text{ (in the binary scale)}$$

Similarly $1011011_2 = 1.2^6 + 1.2^4 + 1.2^3 + 1.2 + 1$ (scale of 10)
$$= 64 + 16 + 8 + 2 + 1$$
$$= 91,$$

using the suffix notation proposed on page 50 for numbers not in the denary scale.

    The complete binary addition and multiplication tables are

$1+0 = 1$	$1 \times 0 = 0$
$1+1 = 10$	$1 \times 1 = 1$

and, at first glance, appear to be the schoolboy's dream of arithmetical heaven. The table of comparison between binary and denary scales above, however, indicates the formidable size of the binary number representing 8 in our decimal scale, for $8 = 1000_2$, four digits being needed. Even more clearly is this point brought out by the second example, where

$$91 = 1011011_2, \text{ requiring 7 digits.}$$

Example. Convert 985 to the scale of 2.

```
2)985
2)492+1
2)246+0        Whence
2)123+0                        985 = 1111011001_2,
2) 61+1
2) 30+1        on reading the result upwards round the edge.
2) 15+0
2)  7+1
2)  3+1
    1+1
```

What we have actually done is to find the multipliers, 0 or 1, of all powers of 2 up to 2^9, the highest needed. In full, we mean that

$$9.10^2+8.10+5 = 1.2^9+1.2^8+1.2^7+1.2^6+0.2^5+1.2^4$$
$$+1.2^3+0.2^2+0.2+1.$$

To convert the opposite way is equally easy.

Example. Express 10111011_2 as an ordinary (denary) number.
Writing this downwards with the *smallest* unit first:

Binary Digit	Multiplier	Product
1	1	1
1	2	2
0	2^2	0
1	2^3	8
1	2^4	16
1	2^5	32
0	2^6	0
1	2^7	128
		187

whence

$$10111011_2 = 187.$$

Obviously it is not necessary to write all this, once the powers of 2 are clearly grasped. More conveniently, a number of moderate size can be converted to the denary scale as above, but omitting the middle column, or can be evaluated thus:

$$10111011_2 = 2^7 + 2^5 + 2^4 + 2^3 + 2 + 1$$
$$= 128 + 32 + 16 + 8 + 2 + 1$$
$$= 187.$$

These examples will serve to illustrate the tiresome property of the binary scale. The average man might well lose his sanity when operating with several large numbers expressed in the scale of 2, but the electronic computer is not in the least bothered. Operations on such a machine are carried out so rapidly that the size of a binary number is of little consequence. It is nevertheless instructive—and, indeed, important to a student of modern mathematics—to carry out simple exercises.

Example. Add the following binary numbers

$$1011, 1101, 111.$$

We have

$$
\begin{array}{r}
1011 \\
1101 \\
111 \\
\hline
11111
\end{array}
$$

Steps: Adding the units column, $1 + 1 + 1 = 11$. Put down 1 and carry 1.

Adding the two's column, $1 + 0 + 1 + 1$ (carry) $= 11$. Put down 1 and carry 1.

Adding the $(two)^2$ (i.e. four's) column, $0 + 1 + 1 + 1$ (carry) $= 11$. Put down 1 and carry 1. Adding the $(two)^3$ (i.e. eight's) column, $1 + 1 + 1$ (carry) $= 11$.

Check: $1011_2 = 11$ $11111_2 = 16 + 8 + 4 + 2 + 1$
         $1101_2 = 13$ $= 31$
         $111_2 = \underline{\ 7\ }$
                  31

Example. Find $10111 - 1101 - 1001$, all the numbers being in the binary system. Check the result by converting all the numbers to the denary system.

We have
$$\begin{array}{r} 10111 \\ -\ \ 1101 \\ \hline 1010 \\ -\ \ 1001 \\ \hline 1 \\ \hline \end{array}$$

i.e. $10111 - 1101 - 1001 = 1$ (in the scale of 2).

[*Check.* $10111_2 - 1101_2 - 1001_2 = 23 - 13 - 9 = 1$

$1_2 = 1$, which tallies.]

Exercise 11

1. Convert the following denary numbers to the binary scale:
   (*a*) 12,   (*b*) 23,   (*c*) 72,   (*d*) 191,   (*e*) 947,   (*f*) 3054,   (*g*) 7^4.

2. Convert the following binary numbers to the denary scale:
   (*a*) 111,   (*b*) 1010,   (*c*) 1111,   (*d*) 11010,   (*e*) 101101,   (*f*) 10110111,
   (*g*) 11100100,   (*h*) 1011100101.

3. Find the following, all the numbers being in the scale of 2,
   (*a*) $101 + 111$,   (*b*) $1011 + 1101$,   (*c*) $111 + 1101$,
   (*d*) $101011 + 110010$,   (*e*) $1011 + 10111 + 1101$,
   (*f*) $11010111 + 10111101$.

4. Subtract 101111_2 from 111011_2.
5. Find the following, all the numbers being in the binary scale,
   (*a*) $1001 - 101 + 11$,   (*b*) $10110 + 101 - 11000$,
   (*c*) $11011 - 110011 + 10111 + 1$,   (*d*) $110111 - (10110 + 10011)$.

6. Convert all the numbers in Qns. 3, 4, 5 above to the denary scale and check the results obtained previously.

Example. In the scale of 2, multiply 111 by 101.

We have
$$\begin{array}{r} 111 \\ 101\ \times \\ \hline 111 \\ 111 \\ \hline 100011 \\ \hline \end{array}$$

i.e. $110 \times 101 = 100011$ (in the scale of 2).

Example. Multiply 101101_2 by 11011_2.

We have

$$
\begin{array}{r}
101101 \\
11011 \times \\
\hline
101101 \\
101101 \\
101101 \\
101101 \\
\hline
10010111111
\end{array}
$$
(in the scale of 2).

Example. Divide 110000111 by 10001, in the scale of 2. Convert the numbers to the denary scale and check the result by division in this system.

$$
\begin{array}{r}
10001)110000111(10111 \text{ (scale of 2)} \\
10001 \\
\hline
11101 \\
10001 \\
\hline
11001 \\
10001 \\
\hline
10001 \\
10001 \\
\hline
\cdots\cdots
\end{array}
$$

[*Check.* $110000111_2 = 1+2+4+128+256 = 391$
$10001_2 = 1+16 = 17$

$$
\begin{array}{r}
17)391(23 \\
34 \\
\hline
51 \\
51 \\
\hline
\cdot\,\cdot
\end{array}
$$

Finally, $10111_2 = 1+2+4+16 = 23$, which tallies.]

Example. Solve, in the binary scale, the equation

$$111x + 1010 = 101101.$$

We have $111x = 101101 - 1010$
$= 100011$

$\therefore\ x = \dfrac{100011}{111}$

$$
\begin{array}{r}
101101 \\
- \quad 1010 \\
\hline
111)100011(101 \\
111 \\
\hline
111 \\
111 \\
\hline
\cdots
\end{array}
$$

$= 101$ (scale of 2).

Exercise 12

1. In the binary scale, find

       (*a*) 101×111,   (*b*) 11×1011,

       (*c*) 110×1101,   (*d*) 10110×1101,

       (*e*) 1111×111,   (*f*) 10111×110110.

2. Find, in the scale of 2,

       (*a*) $1100 \div 10$,   (*b*) $10101 \div 11$,

       (*c*) $1101001 \div 101$.

3. Solve the binary scale equations

       (*a*) $11x = 1001$,   (*b*) $1011x = 110111$,

       (*c*) $111x + 1001 = 11110$.

4. Solve, in the binary system, the simultaneous equations

$$101x - 10y = 1000,$$

$$11x + 100y = 1010.$$

[*Hint*. Multiply the first equation by 10 and add to the second.]

5. Convert the following binary numbers to the denary system:

    (*a*) 101011, (*b*) 1110010110, (*c*) 110110010011101.

6. Find, in the binary system,

    (*a*) $10110 + 100101 + 11101$, (*b*) $110000101 - 101111111$,

    (*c*) $11011 - 111011 + 110111$.

7. Evaluate

    (*a*) $\dfrac{11100 \times 1111}{1100}$,   (*b*) $\dfrac{11110 \times 101101}{11001}$.

8. Divide, in the scale of 2,

       (*a*) 1000010 by 1011, (*b*) 101101001 by 10011,

       (*c*) 10000110001 by 11101.

9. Find $\sqrt{101101001_2}$. [*Hint*. Consider 8(*b*) above *or* convert to the denary scale, find the square root and convert back to the binary scale.]

10. Check the results in Qns. 3, 4, 6–8 above by converting to the denary scale and carrying out the processes indicated.

4. Bicimals (Binary Fractions)

We are familiar with the decimal point. The number $3 \cdot 1$, for example, indicates $3\frac{1}{10}$, the $0 \cdot 1$ standing for $\frac{1}{10}$. Similarly, $4 \cdot 55$ indicates $4\frac{55}{100}$, which can be reduced to $4\frac{11}{20}$.

In the binary scale, therefore, 0.1_2 should analogously mean $\frac{1}{2}$. This is clear if we look at, say, 11_2; here the one in the "unit's" column is one-half of the one in the "two's" column (for $11_2 = 2+1$).

Thus, $$11.1 \text{ must mean } 2+1+\tfrac{1}{2} = 3\tfrac{1}{2}.$$

Similarly, $$101.1_2 = 4+0+1+\tfrac{1}{2} = 5\tfrac{1}{2}.$$

Extending this, what meaning do we assign to 0.01_2? A moment's reflection suggests 0.01_2 is one-half of 0.1_2, i.e. $0.01_2 = \frac{1}{4}$ (denary scale).

The point in such a number as 101.1101_2 can reasonably be called the *bicimal point* and the digits following it form a bicimal fraction.

The whole binary system is therefore

2^2	2^1	2^0	2^{-1}	2^{-2}
fours	twos	ones	halves	quarters

[Note that $2^{-1} = \frac{1}{2}$, $2^{-2} = \frac{1}{4}$, and so on.]

It follows that any bicimal can be expressed as a denary vulgar fraction and hence also in decimal form.

Example. Express 101.1101_2 in the scale of 10, (*a*) as a mixed fraction, (*b*) in decimal form.

We have $101.1101_2 = 4+1+\frac{1}{2}+\frac{1}{4}+\frac{1}{16}$
$$= 5\tfrac{13}{16} \text{ (mixed fraction)}$$
$$= 5.8125 \text{ (decimal form).}$$

Let us now consider how we would convert a recurring *bicimal* to the scale of 10. It is first necessary to think of the meaning we assign to recurring *decimals*.

We have $0.9999\ldots = 1$, for by proceeding with our recurring decimal on the left-hand side, we get as near to 1 as we please, i.e.

$$0.\dot{9} = 1. \text{ It follows that } 0.\dot{1} = \frac{0.111\ldots}{1} = \frac{0.111\ldots}{0.999\ldots} = \frac{111\ldots}{999\ldots} = \frac{1}{9},$$

for the top is one-ninth of the bottom. Similarly we get

$$0.\dot{2} = \frac{2}{9}; \quad 0.\dot{3} = \frac{3}{9} = \frac{1}{3} \quad \text{and so on.}$$

We can extend the idea, thus:

$$0.\dot{2}\dot{4} = \frac{24}{99} = \frac{8}{33}; \quad 0.\dot{2}7\dot{9} = \frac{279}{999} = \frac{31}{111}.$$

This gives us the clue to recurring bicimals, for by analogy we get

$$0 \cdot \dot{1}_2 = 0 \cdot 1111 \ldots = 1 \text{ (scale of 2)} = 1 \text{ (scale of 10)}$$

$$0 \cdot \dot{0}\dot{1}_2 = \frac{1}{11} \text{ (scale of 2)} = \tfrac{1}{3} \text{ (scale of 10)}$$

$$0 \cdot \dot{1}1\dot{0}_2 = \frac{110}{111} \text{ (scale of 2)} = \frac{6}{7} \text{ (scale of 10)}$$

Example. Express $0 \cdot 0\dot{1}0\dot{1}_2$ as a fraction   (*a*) in the scale of 2,   (*b*) in the scale of 10.

We firstly observe that the first zero does not recur. The given binary number is in fact $0 \cdot 0101101101 \ldots$

By extending the working above, $0 \cdot \dot{1}0\dot{1}_2 = \frac{101}{111}$ (scale of 2) and to

move the bicimal point one place to the left merely means that we . need to divide by 2 (scale of 10), i.e. by 10 (scale of 2),

$$\Rightarrow \quad 0 \cdot 0\dot{1}0\dot{1}_2 = \frac{1}{10} \times \frac{101}{111} = \frac{101}{1110} \text{ (scale of 2)},$$

$$= \frac{5}{14} \text{ (scale of 10)}.$$

Exercise 13

1. Express the following bicimal numbers (i) as vulgar or mixed fractions in the scale of 10, (ii) in decimal form:

(*a*) 0·11,   (*b*) 0·0101,   (*c*) 0·10011,   (*d*) 101·01,   (*e*) 110·1101,

(*f*) 0·0$\dot{1}$,   (*g*) 1001·1$\dot{0}$,   (*h*) 0·0$\dot{0}\dot{1}$,   (*i*) 0·$\dot{0}$1$\dot{1}$,   (*j*) 0·$\dot{1}$00$\dot{1}$.

The conversion of denary fractions to bicimals requires a little more care, as more often than not the bicimal proves to be recurring. We have seen that $\tfrac{1}{2} = 0 \cdot 1_2$ and that $\tfrac{1}{4} = 0 \cdot 01_2$, but $\tfrac{1}{3}$ (shown above, in reverse) and $\tfrac{1}{5}$, for example, require a little thought.

We start by converting $\tfrac{1}{5}$ into a bicimal vulgar fraction. Now

$5 = 101_2$, hence $\frac{1}{5} = \frac{1}{101}$ (scale of 2). Long division in the binary scale gives

$$
\begin{array}{r}
101)\overline{1 \cdot 0000000}(0 \cdot 00110011 \ldots \\
\underline{101} \\
110 \\
\underline{101} \\
1000 \\
\underline{101} \\
110 \\
\underline{101} \\
\text{etc.}
\end{array}
$$

Hence $\dfrac{1}{5} = 0 \cdot 0\dot{0}1\dot{1}$ (scale of 2).

[Those well versed in binary arithmetic may observe that $15 = 1111_2$, and so

$$\frac{1}{5} = \frac{3}{15} = \frac{11}{1111} \text{ (scale of 2)} = 0 \cdot 0\dot{0}1\dot{1} \ !]$$

A few results are listed below

Denary Fraction	Bicimal	Denary Fraction	Bicimal
$\dfrac{1}{3}$	$0 \cdot \dot{0}\dot{1}$	$\dfrac{1}{5}$	$0 \cdot 0\dot{0}1\dot{1}$
$\dfrac{1}{7}$	$0 \cdot \dot{0}0\dot{1}$	$\dfrac{1}{9}$	$0 \cdot \dot{0}0011\dot{1}$
$\dfrac{1}{15}$	$0 \cdot \dot{0}00\dot{1}$	$\dfrac{1}{17}$	$0 \cdot \dot{0}0001111\dot{1}$

It is easy to extend these results, e.g.,

$$\frac{3}{7} = 3 \times \frac{1}{7} \Rightarrow 11 \times \dot{0} \cdot 00\dot{1}_2 = 0 \cdot \dot{0}1\dot{1}_2$$

$$\frac{1}{6} = \frac{1}{2} \times \frac{1}{3} \Rightarrow 0 \cdot 1 \times 0 \cdot \dot{0}\dot{1} = 0 \cdot 0\dot{0}\dot{1}_2$$

[Note that $0 \cdot 00\dot{1}_2 = 0 \cdot 001010101 \ldots$ in which the first zero after the bicimal point does not recur.]

The conversion of decimals to bicimals is naturally linked to the conversion of denary integers to binary integers but there is a fundamental difference. In converting denary integers to binary, as earlier in this chapter, it will be recalled that we *divided*[9] repeatedly by 2; but in converting decimals to bicimals we *multiply* repeatedly by 2. The following example illustrates the method.

Example. Convert 0·59375 to a bicimal.

0	·59375 × 2
1	·1875
0	·375
0	·75
1	·5
1	·0

The result is
read downwards
in the left hand
column

$$0 \cdot 59375 = 0 \cdot 10011_2.$$

The bicimal obtained above terminates, but many do not.

Example. Convert 0·7 to a bicimal.

0	·7 × 2
1	·4
0	·8
1	·6
1	·2
0	·4
0	·8
1	·6
1	·2
0	·4

These figures recur

Hence $0 \cdot 7 = 0 \cdot 1\dot{0}11\dot{0}_2$, where the first 1 after the bicimal point does *not* recur (in the calculation, 1·4 [line 2] and 0·4 [line 6] are not the same).

[9] Page 64.

Exercise 14

1. Convert the following vulgar fractions to bicimals, recurring where necessary:

$$(a)\ \tfrac{3}{8}, \quad (b)\ \tfrac{11}{16}, \quad (c)\ \tfrac{2}{7}, \quad (d)\ \tfrac{5}{9}, \quad (e)\ \tfrac{14}{15}, \quad (f)\ \tfrac{7}{12}.$$

2. Convert the following decimals to exact bicimals:

$$(a)\ 0.625, \quad (b)\ 0.4375, \quad (c)\ 0.15625, \quad (d)\ 0.171875.$$

3. Convert the following decimals to recurring bicimals:

$$(a)\ 0.8, \quad (b)\ 0.6, \quad (c)\ 0.9, \quad (d)\ 0.85, \quad (e)\ 0.\dot{6}, \quad (f)\ 0.\dot{8}, \quad (g)\ 0.9\dot{3}.$$

In practice other methods of conversion are sometimes used for bicimals, but those above are adequate for a grasp of the principles.

5. The Octal Scale

It is obviously tedious to carry out many repeated operations in which multiplication or division by 2 is involved. The octal scale (scale of 8) offers a conveniently short cut. We convert from the denary system to the octal and then very simply to the binary.

In the octal scale, we use the digits 0, 1, 2, 3, 4, 5, 6, 7 only, and as $7_8 = 1.2^2 + 1.2 + 1$, then $7_8 = 111_2$. We observe, however, that up to the number 7 the denary and octal scales are the same:

Denary *or* Octal	1	2	3	4	5	6	7
Binary	001	010	011	100	101	110	111

In the table above we have deliberately put in zeros when apparently unnecessary for the denary (octal) numbers 1, 2, 3, thereby using three binary digits, for a reason shortly to be explained.

The procedure is illustrated by an example:

Example. Convert 3407 to the binary scale.

```
8)3407        |
8) 425+7      |
8)  53+1     ↑
     6+5     |
          →
```

Reading the result round the edge

$$3407 = 6517_8$$

In full, the result will read $6.8^3 + 5.8^2 + 1.8 + 7$ and it can be laid out as:

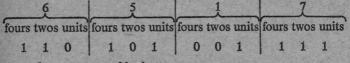

6			5			1			7		
fours	twos	units	fours	twos	units	fours	twos	units	fours	twos	units
1	1	0	1	0	1	0	0	1	1	1	1

using the conversion table above.

$$\therefore \quad 3407 = 6517_8 = 110101001111_2.$$

Note. We cannot proceed *directly*, in this way, from the denary scale to the binary, because the extra digits 8 and 9 in the scale of ten convert into 1000_2 and 1001_2 respectively. These require four digits but, unfortunately, do not exhaust all the four digit possibilities and in consequence would lead to an invalid method.

Exercise 15

1. Convert the following denary numbers to the octal scale:

     (a) 79,   (b) 148,   (c) 476,   (d) 3918,

     (e) 10725,   (f) 24938,   (g) 79924.

2. Convert the following numbers to the binary scale:

     (a) 46_8,   (b) 74_8,   (c) 423_8,   (d) 576_8,

     (e) 3040_8,   (f) 6732_8,   (g) 20723_8.

3. Convert the following denary numbers to the binary scale by first converting to the octal scale as in the example above this set of exercises:

     (a) 218,   (b) 1309,   (c) 2790,

     (d) 4683,   (e) 8117,   (f) 8496.

4. Convert the following numbers to the denary scale:

     (a) 431_8,   (b) 432_6,   (c) 101101_2,

     (d) 211012_3,   (e) 7185_9,   (f) 40231_5.

5. Convert the following denary numbers to the scales indicated:

     (a) 237 to scale 8,   (b) 1465 to scale 5,

     (c) 1932 to scale 9,   (d) 6160 to scale 4.

6. Construct a multiplication table in the scale of 8, in the same way as that for the scale of 6, shown on page 62. Use the table to make the following calculations in the scale of 8:

    (a) 35×14,   (b) 42×31,   (c) 102×45,

    (d) $44 \div 6$,   (e) $61 \div 7$,   (f) $526 \div 6$.

7. Find the following in the scales indicated:

    (a) $215 + 372 - 134$ (scale of 8),

    (b) $2101 - 1022$ (scale of 3),

    (c) $4152 - 3125 + 1034$ (scale of 6).

8. Find, in decimal form, the following:

    (a) $0 \cdot 1001$ (scale of 2),   (b) $0 \cdot 1\dot{0}$ (scale of 2),

    (c) $0 \cdot 1$ (scale of 3),   (d) $0 \cdot 11$ (scale of 3),

    (e) $0 \cdot \dot{1}0\dot{2}$ (scale of 3),   (f) $0 \cdot 54$ (scale of 8),

    (g) $0 \cdot \dot{6}\dot{1}$ (scale of 8).

CHAPTER 5

MODULAR ARITHMETIC

1. Modulo Systems

Suppose we consider the measurement of time. In the 12-hour clock, once the hour hand has travelled past the 12 mark, counting starts again. If, for example, the hour hand starts at the 12 o'clock position and moves 15 hours, it finishes at 3 o'clock. The multiple of 12 has been removed and only the remainder is used; thus

$$15 = 1.12 + 3.$$

The same procedure is adopted after 2 or more rotations, e.g. 39 hours after starting we again find the hour hand at 3 o'clock, for

$$39 = 3.12 + 3.$$

We can express this as

$$39 \equiv 3 \pmod{12},$$

which is read "39 is congruent to 3, modulo 12".

In general, two numbers are congruent in any modulo system if, after the removal of whole number multiples of the modulo, the remainders are equal. Only integers are used in the process, which is the result of the work of Karl Friedrich Gauss (1777–1855), who published his findings on the theory of numbers in *Disquisitiones Arithmeticae* (1801).

It is obvious that the remainder must be zero or a positive integer less than the modulo. Thus, if the modulo is 7, say, the remainder is 0, 1, 2, 3, 4, 5 or 6. More generally, for modulo n, the remainder is $0, 1, 2 \ldots (n-2)$ or $(n-1)$.

Let us examine the remainders when the set of all integers, including zero, is divided by 4. We have the following situation, where the remainders 0, 1, 2, 3, are repeated over and over again.

$$\begin{array}{lll}
\cdot\;\cdot\;\cdot\;\cdot\;\cdot\;\cdot\;\cdot\;\cdot\;\cdot & 0 = 0.4+0 & 6 = 1.4+2 \\
-5 = -2.4+3 & 1 = 0.4+1 & 7 = 1.4+3 \\
-4 = -1.4+0 & 2 = 0.4+2 & 8 = 2.4+0 \\
-3 = -1.4+1 & 3 = 0.4+3 & 9 = 2.4+1 \\
-2 = -1.4+2 & 4 = 1.4+0 & 10 = 2.4+2 \\
-1 = -1.4+3 & 5 = 1.4+1 & \cdot\;\cdot\;\cdot\;\cdot\;\cdot\;\cdot\;\cdot\;\cdot
\end{array}$$

From this it will be seen, for example, that

$$-5 \equiv -1 \equiv 7 \equiv 3 \pmod 4.$$

Suppose now we carry out some simple addition in the modulo 4 system, e.g.

$$7+8+10 = 25 \equiv 1 \pmod 4;$$

but we could have used the table above, thus,

$$7+8+10 \equiv 3+0+2 = 5 \equiv 1 \pmod 4.$$

This method could save a certain amount of arithmetic in the case of large numbers, by dividing each number mentally by 4 (or whatever modulo we are using), replacing each by its remainder, adding the remainders and repeating the process if necessary.

Example. Find the remainder for

$$317+584-2193 \pmod 5$$

Dividing each in turn mentally by 5, we have

$$317+584-2193 \equiv 2+4-3 = 3 \pmod 4$$

[What we do is to take $317 \div 5 =$ something $+$ remainder 2, where we do not even bother to remember the "something" as we divide out "in our heads". Actually $317 \div 5 = 63.5+2$.]

For interest we now give two tables in modulo 4:

(a) *Addition Table*

+	1	2	3	4...
1	2	3	0	1
2	3	0	1	2
3	0	1	2	3
4	1	2	3	0

From the table, for example, $3+2 \equiv 1 \pmod 4$.

(b) Multiplication Table

×	1	2	3	4...
1	1	2	3	0
2	2	0	2	0
3	3	$\boxed{2}$	1	0
4	0	0	0	0

From the table, for example, $2 \times 3 \equiv 2$ (mod 4).

Example. Find the remainder for 21×39 (mod 4).
   We have $21 \times 39 = 819 \equiv 3$ (mod 4).
   We may also use the short cut we applied for addition and subtraction in the previous example, thus

$$21 \times 39 \equiv 1 \times 3 \equiv 3 \text{ (mod 4)}.$$

This is far from obvious and a proof is given for those interested. [The reader who views algebra with a jaundiced eye can omit the following theorem without handicap.

Theorem. If $p \equiv a$ and $q \equiv b$ (mod n), then $pq \equiv ab$ (mod n).
   Proof. If $p \equiv a$ (mod n), then $p = xn + a$
      and if $q \equiv b$ (mod n), then $q = yn + b$,
all the letters representing integers. (Whatever integers x and y are is immaterial).
   Then $pq = (xn + a)(yn + b)$
$$= xyn^2 + ayn + bxn + ab$$
$$= n(xyn + ay + bx) + ab$$
$$\equiv ab \text{ (mod } n),$$
for clearly n divides into $n(xyn + ay + bx)$ exactly.]

Example. Construct the multiplication table for 7 in modulo 5 system. We have

$1 \times 7 \equiv 2$	$4 \times 7 \equiv 3$	$7 \times 7 \equiv 4$
$2 \times 7 \equiv 4$	$5 \times 7 \equiv 0$	etc.
$3 \times 7 \equiv 1$	$6 \times 7 \equiv 2$	in modulo 5.

There is something of a danger, at this stage, that one may be tempted to discard the elementary arithmetic tables normally learned in primary school. This would be a serious mistake, for it only requires a moment's reflection to observe that throughout this work we are having recourse to mental addition and multiplication in the denary scale, even although our answers are not necessarily in this system.

Example. Find which of the following numbers are congruent, modulo 7:

$$(a)\ 213,\quad (b)\ 459,\quad (e)\ 605.$$

We have
$$213 = 30.7 + 3 \equiv 3 \ (\text{mod } 7)$$
$$459 = 65.7 + 4 \equiv 4 \ (\text{mod } 7)$$
$$605 = 86.7 + 3 \equiv 3 \ (\text{mod } 7)$$
$$\therefore\ 213 \equiv 605 \ (\text{mod } 7)$$

Exercise 16

1. Write down addition and multiplication tables for the numbers 1, 2, 3, 4, 5 in modulo 6.

2. Write down the multiplication table for 7 in (a) modulo 8, (b) modulo 10.

3. Show that   (a) $316 \equiv 7240 \ (\text{mod } 6)$,   (b) $-500 \equiv 900 \ (\text{mod } 7)$.

4. Find the smallest positive integer (or zero, when applicable) congruent with the given number for the given modulo in each of the following cases:

(a) 2158 (mod 5),   (b) 32774 (mod 17),   (c) -5 (mod 9),
(d) -394 (mod 6),   (e) $16 + 9 - 14 - 20$ (mod 3),
(f) $372 - 425$ (mod 7),   (g) 34×62 (mod 5),   (h) 694×217 (mod 6),
(i) 694×217 (mod 7),   (j) -61×38 (mod 11).

2. Algebraic Congruences

The methods used in the section above can be applied to some important theorems in arithmetic, but they are rather beyond the scope of this book. We can, however, easily solve certain types of algebraic problem.

Example. Solve, in integers, the congruence $2x+1 \equiv 4 \pmod 5$.

We have $\hspace{3.5cm} 2x \equiv 3 \pmod 5$.

Multiply both sides by 3,

$$\therefore \; 6x \equiv 9 \pmod 5.$$

Now the left hand side is $5x+x$, but $5x \equiv 0 \pmod 5$, because 5 times any integer is exactly divisible by 5 and leaves remainder 0.

$$\therefore \; 5x+x \equiv x \pmod 5$$

Hence $\hspace{3.5cm} x \equiv 9 \equiv 4 \pmod 5,$

and the complete solution is

$$x = 5n+4$$

where n is any integer (positive, negative or zero).

[*Check.* $2x+1 = 10n+8+1 = 10n+9 \equiv 9 \equiv 4 \pmod 5$.]

Exercise 17

Solve the following congruences, Qns. 1–4, giving the general solution in integers:

1. $x+3 \equiv 0 \pmod 4$.
2. $2x-1 \equiv 5 \pmod 7$.
3. $5x+6 \equiv 0 \pmod 8$. [*Hint.* Replace 0 by 8, for $8 \equiv 0 \pmod 8$.]
4. $6x+4 \equiv 2x-3 \pmod 5$.
5. Find the three next smallest positive integers congruent with 6, modulo '17.
6. Why is there no solution in integers to the congruence $3x \equiv 1 \pmod 6$?

[*Hint.* We see this if we draw up part of the multiplication table for 3, mod 6.

×	1	2	3	4	5	6
1	1	2	3	4	5	0
2	2	4	0	2	4	0
→ 3	3	0	3	0	3	0

for no number multiplied by 3 is congruent to 1 (mod 6) in the table. In fact, $3x$ can only be congruent to 0 or 3 for this modulus, for $3.1 \equiv 3$, $3.2 \equiv 0$, $3.3 \equiv 3$, $3.4 \equiv 0$, $3.5 \equiv 3, \ldots \pmod 6$.]

7. Is there a solution, in integers, to $4x+1 \equiv 7 \pmod 5$? If there is, find it. If not, explain why it does not exist.

3. Tests for Factors of Whole Numbers

We give below a series of tests, together with proofs that all are valid. Some are well known and easily proved but others are interesting in the application of methods kindred to those in earlier sections of this chapter. Altogether, the tests are sufficient for finding all the factors from 2 to 16 inclusive of any integer. The first factor, therefore, for which a test is not given is 17. The test for divisibility by 7, 11 or 13 is of little value unless the number under consideration is large. We group the tests for convenience of proof.

We use the (non-standard) abbreviation *div* for "is divisible by" and we adopt X as the integer we are investigating. We also introduce the symbol $\Leftrightarrow$ meaning "implies and is implied by", or "if and only if".

(a) X *div* 2, 4, 8, 16.

The number X is divisible by 2, 4, 8, 16, if the number formed by the last 1, 2, 3, 4 digits of X is divisible by 2, 4, 8, 16 respectively.

Let us consider 213648.
Here $X = 213648 = 213 \times 1000 + 648$.
Now 1000 *div* 8 (for $1000 = 8 \times 125$).
so X *div* $8 \Leftrightarrow 648$ *div* 8, but $648 = 8 \times 81$,

$$\therefore \quad X \text{ div } 8.$$

The general proof is equally simple. We take the case of division by 4, but 2, 8, 16 are proved exactly similarly. (16 is, however, rarely used.)
Any number $X = A.100 + B$, where A, B are integers and B is less than 100.

$$\therefore \quad X \text{ div } 4 \Leftrightarrow B \text{ div } 4 \text{ (for 100 div 4).}$$

More generally, $X \equiv B \pmod 4$, it is of interest to observe.

Example. Is 43729536 divisible by 8?

$$536 \text{ div } 8 \text{ (for } 536 \div 8 = 67)$$

$$\therefore \quad 53729536 \text{ div } 8.$$

(b) X *div* 3 or 9.

$$X = a_0 + a_1.10 + a_2.10^2 + \ldots + a_n.10^n$$

where[10] every $a_r \in \{0, 1, 2, \ldots 9\}$, for $r = 0, 1, 2 \ldots n$.

[Actually X would look like $a_n, a_{n-1}, a_{n-2}, \ldots a_3, a_2, a_1, a_0$, e.g. $X = 21468$ has $a_0 = 8, a_1 = 6, a_2 = 4, a_3 = 1, a_4 = 2$.]

Let $P = a_0 + a_1 + a_2 + \ldots + a_n$

$\therefore$ $X - P = 9a_1 + 99a_2 + 999a_3 + \ldots + 99\ldots9a_n$

$= 9k$ (say, where k is an integer).

If P div 3, $P = 3q$ (q an integer)

$\therefore$ $X = 3q + 9k = 3(q + 3k)$

$\Rightarrow X$ div 3.

Similarly, if P div 9, $P = 9r$ (r an integer)

$\therefore$ $X = 9r + 9k = 9(r + k)$

$\Rightarrow X$ div 9.

[*Note*. We can lay out the second part of this proof in congruence form, thus:

$$X - P = 9k \text{ (k an integer)}$$

$\therefore$ $X - P \equiv 0 \pmod 3$ and $X - P \equiv 0 \pmod 9$

$\therefore$ X div $3 \Leftrightarrow P$ div 3   and X div $9 \Leftrightarrow P$ div 9.]

Hence we have the theorem: An integer X is divisible by 3 or 9 if the sum of its digits is divisible by 3 or 9 respectively.

Example. Is 45738 divisible by 3 or by 9?

We have $4 + 5 + 7 + 3 + 8 = 27$

and $27 \div 9 = 3$

$\therefore$ 45738 div 9 (and hence div 3).

(c) X div 6 or 12

X div 6 if div 2 and div 3 above.

X div 12 if div 4 and div 3 above.

Example. Is 28614 divisible by 6 or by 12?

$$2 + 8 + 6 + 1 + 4 = 21 \ (div \ 3).$$

Also the number 28614 is even (div 2)

but $14 \div 4$ is not an integer

$\therefore$ 28614 div 6 but not div 12.

[10] Fully explained on pp. 86–88. Alternatively, we can say every a_r is a positive integer less than 9 or is zero, for any value of r from 0 to n.

(d) X *div* 5 or 10

Again $X = a_0 + a_1.10 + a_2.10^2 + \ldots + a_n.10^n$

$\therefore \quad X = a_0 + 10m$ (where m is an integer).

If a_0 *div* 5, $a_0 = 5s$ (s an integer)

$$\therefore \quad X = 5(s + 2m)$$
$$\text{i.e. } X \text{ div } 5.$$

If $a_0 = 0$, $X = 10m$

$$\therefore \quad X \text{ div } 10.$$

Hence we have the theorem: An integer X is divisible by 5 or 10 if it ends in 5 or 0 respectively.

(e) X *div* 15

$$X \text{ div } 15 \text{ if } \text{div } 3 \text{ and } \text{div } 5 \text{ above.}$$

Example. Is 24735 divisible by 15?

$$X = 24735 \text{ div } 5 \text{ (for it ends in 5)}.$$
$$\text{Also } 2 + 4 + 7 + 3 + 5 = 21 \ (\text{div } 3)$$
$$\therefore \quad 24735 \text{ div } 15.$$

(f) X *div* 7, 11, 13

This method is more subtle[11] and the proof requires a little fuller appreciation of algebra. We write

$$X = p_0 + p_1.10^3 + p_2.10^6 + p_3.10^9 + \ldots + p_n.10^{3n}$$

where $p_0, p_1, p_2 \ldots p_{n-1}$ each have three digits (although some may be zero) and p_n has three digits or less.

(For example,

$$4729080642 = 642 + 80.10^3 + 729.10^6 + 4.10^9.$$

We do not need to write 080.10^3.)

Let $Y = p_0 - p_1 + p_2 - p_3 + \ldots + (-1)^n p_n$

$\therefore \quad X - Y = p_1(10^3 + 1) + p_2(10^6 - 1) + p_3(10^9 + 1) + \ldots$
$$+ p_n(10^{3n} + [-1]^n).$$

Now from elementary algebra (we can use the remainder theorem) all the factors

$$10^3 + 1, \ 10^6 - 1, \ 10^9 + 1, \ldots 10^{3n} + [-1]^n$$

---

[11] Some readers may prefer to omit the *proof* and just read the first few lines. The application of the method is shown in the example immediately following.

of terms in the right hand side themselves have the common factor $10^3 + 1$.

But $10^3 + 1 = 1001 = 7 \times 11 \times 13$.

$$\therefore \quad X - Y = 7.11.13.Z \text{ (where } Z \text{ is an integer)}$$

$$\therefore \quad X \ div \ 7, 11, 13 \Leftrightarrow Y \ div \ 7; 11, 13$$

[Alternatively we can say $X \equiv Y \pmod{7, 11, 13}$ so if $Y \equiv O$, then $X \equiv 0 \pmod{7, 11, 13}$.]

This is clear if we consider one case.

Suppose $Y = 7u$, where u is an integer,

then $\qquad X = 7u + 7.11.13.Z$

$$= 7 (u + 11.13.Z)$$

$$\therefore Y \ div \ 7 \Rightarrow X \ div \ 7.$$

Example. Find whether 7, 11 or 13 are factors of 385086481.

Here $p_0 = 481, p_1 = 86, p_2 = 385$

$\therefore \quad Y = p_0 - p_1 + p_2 = 481 - 86 + 385$

$$= 780, div \ 13, \text{ not } div \ 7, 11.$$

$\therefore \quad 385086481 \ div \ 13, \text{ not } div \ 7, 11.$

(*g*) $X \ div \ 14$.

$$X \ div \ 14 \text{ if even (i.e. } div \ 2) \text{ and } div \ 7.$$

Example. Find whether 14 is a factor of 3250044.

Here $p_0 = 44, p_1 = 250, p_2 = 3$

$\therefore \quad Y = p_0 - p_1 + p_2 = 44 - 250 + 3$

$$= -203, div \ 7.$$

But 3250044 is even and divisible by 7, therefore it is divisible by 14.

Exercise 18

1. Find all the factors of the following numbers:
        (*a*) 60,   (*b*) 84,   (*c*) 385.

2. Find the prime factors (i.e. factors which are prime numbers) of
       (a) 1170, (b) 2431, (c) 37401.

3. Find which of the following numbers are divisible by 15:
       (a) 27295, (b) 38415, (c) 263570, (d) 10914.

4. Find whether the following numbers are divisible by 7, 11 or 13:
       (a) 216425, (b) 81292315, (c) 513480467.

5. Find all the factors, which are less than 16, of 14730765.

6. Find all the factors less than 16 of 292564187097.

7. Prove that $20746864 \equiv 138 \pmod{11, 13, 14}$.

[*Hint*. It is not necessary to work this out fully. Notice that both of the given numbers are even.]

CHAPTER 6

INTRODUCTION TO SETS

1. Sets

A set is a collection of things considered together in some way. One may have a set of playing-cards, a set of knives, forks and spoons or even a set of children in a school. The reader will recall that, in Chapter 1, we referred to a number line and that we explained that the real numbers could be considered as filling the entire number line. In this sense, the number line consists of the set of points representing the real numbers. More generally, any line can be considered as the set of points filling it.

A cabinet of tableware may contain

6 knives	6 dessert spoons
6 table forks	2 table spoons
6 dessert forks	6 tea spoons.

The total is 32 pieces and constitutes the set in the cabinet. There is, however, the difficulty that, whereas we could distinguish between a table fork and a dessert fork, we might not be able to tell one dessert fork from another.

In mathematics we usually require that the items in a set are distinguishable from one another. Thus a pack of 52 playing-cards would satisfy the criterion, and so would the children in a school, but the cabinet referred to would not, unless each piece were specially marked. There are, however, problems in which this difficulty can be overcome (see, for example, H.C.F. and L.C.M.).

The representation of a set is shown inside { }, which are called *braces*. We do *not* use ( ), which we call *brackets*, for this purpose. For example, {1, 2, 3, 4, 5, 6, 7, 8, 9} is the set of the natural numbers less than 10. Each number is a *member* (element) of the set.

We can use the braces in a slightly different way, e.g.

$$\{d : d \text{ is a playing card}\}$$

which is read as "the set of d things such that d is a playing-card" This is sometimes written $\{d|d \text{ is a playing-card}\}$, where the symbol $:$ *or* $|$ is read "such that", in either case, but only if shown inside braces, otherwise $:$ is merely a colon or a ratio symbol (such as $2:3$, meaning $\frac{2}{3}$)!

2. Subsets

Every set possesses subsets. A subset is a set consisting of none, some or all of the members of a set. These include the set itself and the empty set. Suppose we consider the set

$$X = \{\text{John, Henry, Mary, Anne}\}.$$

Examples of subsets of X are {John, Mary}, {Henry, Mary, Anne} and, of course, {John, Henry, Mary, Anne}.

We use a capital letter to represent a set and we use the symbol $\in$ to mean "is a member of". In the set above, Anne $\in X$, and this is read "Anne is a member of the set X"; also {Anne} is a subset of X.

An empty set is a set with no members and is denoted by ϕ. The subsets of $\{1, 2, 3\}$ are

$$\phi, \{1\}, \{2\}, \{3\}, \{1, 2\}, \{1,3\}, \{2, 3\}, \{1, 2, 3\},$$

i.e. 8 in number. This is 2^3 and one wonders whether there is any significance in the fact that this result is a power of 2. That this is so is easily shown.

Suppose we have a set with n members, then in a subset we may accept or reject any member. The *total* number of ways of doing this is $2 \times 2 \times 2 \times \ldots 2$ (n times), i.e. 2^n. Hence there are 2^n subsets of a set having n members.

The student may find this easier to see if we apply the principle in greater detail to our example of the set $\{1, 2, 3\}$ above:

$$\{\ldots\} \{1 \ldots\} \{.2.\} \{\ldots 3\}$$
$$\{12.\} \{1.3\} \{.23\} \{123\}$$

The dots indicate that we have ignored (i.e. rejected) the members so replaced.

Just as the symbol ∈ means "is a member of" so ∉ means "is not a member of", e.g.

if $A = \{\text{quadrupeds}\}$,

then   Horse $\in A$; Hen $\notin A$; Crocodile $\in A$.

Exercise 19

1. N is the set of all integers. Write in set notation whether the following are members,   (a) 5,   (b) -4,   (c) $1\frac{1}{2}$,   (d) $\sqrt{2}$,   (e) $\sqrt{9}$.
   [*Hint.* $5 \in N$.]

2. R is the set of rational numbers. Write in set notation whether

$$4{\cdot}5, \quad -\frac{3}{7}, \quad \frac{1}{\sqrt{3}}, \quad 0{\cdot}7, \quad \log 4$$

are members.

3. H is the set of hearts in a pack of cards. Are the following members of H, (a) the ace of hearts, (b) an ace?

4. Write down all the subsets of $\{1, 3, 5, 7\}$. How many subsets are there?

5. Write down the set formed from the first seven prime numbers (including 1). How many subsets are there?

6. Are (a) a square, (b) a rhombus, (c) a rectangle, (d) a quadrilateral, members of the set of parallelograms?

7. Is the set of parallelograms a subset of the set of trapeziums?

8. Find which of the numbers 47, 57, 257, 289 are members of the set of prime numbers.

9. If X is the set of the different triangles in Fig. 40, how many subsets are there of X?

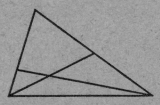

Fig. 40

3. Symbolism

The symbol $\subset$ means "is a subset of", e.g.

$$X \subset Y$$

is read "X is a subset of Y".

Similarly $\supset$ means "contains as a subset", e.g.

$$Y \supset X$$

is read "Y contains X as a subset".

The two statements $X \subset Y$ and $Y \supset X$ are the same and we can write $X \subset Y \Leftrightarrow Y \supset X$, where $\Leftrightarrow$ means "if and only if", as used earlier.

If, however, $X \subset Y$ and $Y \subset X$ (and we must note carefully the direction of the subset symbol), then $X = Y$, for every element of X is an element of Y and, conversely, every element of Y is an element of X and hence X and Y are the same set.

Example. If $X = \{1, 3, 4, 6, 7\}$ and if $A = \{1, 3, 5\}$, $B = \{1, 3, 6\}$, $C = \{1, 3, 7\}$, which of A, B, C are subsets of X?

Clearly $B \subset X$ and $C \subset X$, for every element of B and of C is an element of X, but A is not a subset of X for one member of A (i.e. 5) is not a member of X. We illustrate this diagrammatically below (Fig. 41).

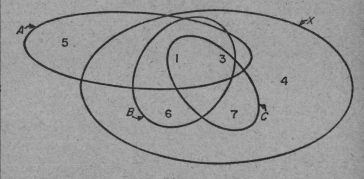

Fig. 41

4. Venn Diagrams

Suppose we draw an oval and put in it the elements of set X of the *example* immediately above.

We now enclose the elements in sets A, B, C in turn by ovals. We see at once that whereas the ovals containing the elements of B and of C can lie entirely inside X, the oval for the elements of A cannot. This leads to the statement that subsets of a given set can be shown enclosed in ovals lying entirely within the set itself. (Strictly speaking, we should use the words "closed curves" in place of "ovals", for some of the shapes obtained may be eccentric!). A figure such as that shown in Fig. 41 is called a *Venn diagram*.

It is often convenient to draw a rectangle around *all* the elements we are considering and to call this set the universal set, $\mathscr{E}$. Thus, again using the *example* above, we would have $\mathscr{E}$ and X as shown in Fig. 42.

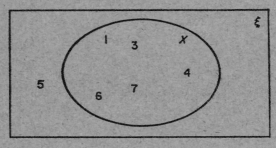

Fig. 42

We do not—and, in fact, cannot—always represent all the individual members of a set on a Venn diagram. Suppose, for example, we wish to show that the set of children in Form 2A is a subset of the second year children in a school, and that this in turn is a subset of the set of all children in the school.

Let $A = \{\text{children in } 2A\}$
                        $B = \{\text{children in second year}\}$
                        $C = \{\text{children in school}\},$
then $A \subset B \subset C$

and, on a Venn diagram, C is the universal set. The diagram appears thus (Fig. 43):

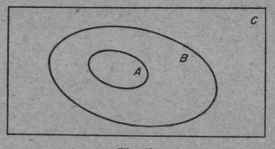

Fig. 43

In this case we do not show any of the members, but we *imagine* them to be represented each by a separate dot. The omission of the members does not cause confusion; the diagram is much clearer without them.

Exercise 20

1. Dogs and cats are animals. Show, on a Venn diagram, that dogs are not cats. [*Hint.* Take $\mathscr{E}$ as the set of all animals.]

2. $\mathscr{E}$ is the universal set of all cards in a pack of 52. If $A = \{\text{diamonds}\}$, $B = \{\text{red cards}\}$, $C = \{\text{queens}\}$, $D = \{\text{court cards}\}$, find (*a*) an actual card x such that $A \supset x$ and $C \supset x$, (*b*) *all* subsets Y of D such that $B \supset Y$ and $C \supset Y$.

3. In question 2 above, would it have been a correct use of notation to have asked the reader to find

　　　(*a*) $\{x : x \in A \text{ and } x \in C\}$,　　(*b*) $\{Y : B \supset Y \text{ and } C \supset Y\}$?

4. $M = \{\text{men}\}$, $W = \{\text{women}\}$, $T = \{\text{tennis-players}\}$ and $\mathscr{E}$ is the set of all people. Show M, W, T in their correct relationship on a Venn diagram. [*Cave.* What about children?]

5. $\mathscr{E}$ is the set of all quadrilaterals, A is the set of squares (non-human), B is the set of rectangles, C is the set of parallelograms, and D is the set of rhombuses. Which of the following statements are true and which are false:

　　(*a*) $C \supset D$,　(*b*) $C \subset B$,　(*c*) $B \supset A$,
　　(*d*) $D \supset A$,　(*e*) $B \supset D \supset A$,　(*f*) $C \supset D \supset A$?

Rewrite the first sentence of this question more briefly in set notation.

6. Explain whether there is a fallacy in the argument:

"A rectangle has four right angles. A certain polygon is known to have four right angles. The polygon is therefore a rectangle." Illustrate the explanation in a Venn diagram.

7. If N = {natural numbers), R = {rational numbers},
I = {integers}, F = {proper fractions}, Q = {real numbers},[12]
draw a Venn diagram showing N, R, I, F, Q, in their correct relationship.

8. Name the fundamental universal set $\mathscr{E}$ of which the following animals are all members: hen, platypus, crocodile, turtle, dinosaur, moth.

5. Intersection of Sets

In the lounge of a home are to be found Mother, Father, Hilary, Peter, the cat and the goldfish. The universal set of animals present is:

$$\mathscr{E} = \{\text{Mother, Father, Hilary, Peter, cat, goldfish}\}.$$

The subset of warm-blooded animals is W, say, where

$$W = \{\text{Mother, Father, Hilary, Peter, cat}\}$$

and the subset of non-human animals is N, say, where

$$N = \{\text{cat, goldfish}\}.$$

Now $\text{cat} \in W$ and $\text{cat} \in N$,
but $N \not\supset W$ and $W \not\supset N$,
so, although W and N have a non-zero subset, to wit, {cat}, in common, neither of the sets N, W is a subset of the other set.

[We use the symbols $\not\supset$ to mean "does not have as a subset" and $\not\subset$ to mean "is not a subset of".]

Using single small letters to represent the elements of $\mathscr{E}$ above (e.g. m stands for Mother) on a Venn diagram, the representation of the subsets W and N is shown in Fig. 44 below.

---

[12] These are all defined in Chapter 1.

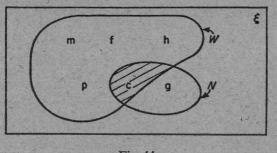

Fig. 44

The time has now come to consider the relationship between, say, set N and set W. Clearly they have something in common (one member, the cat). The sets N and W are said to *intersect*, the symbol used being $\cap$. Thus we have

$$W \cap N = \{cat\}.$$

This is read "The intersection of W and N is the set having one member, the cat." Clearly $\{cat\}$ is a subset of W and of N.

Definition. The intersection of two sets A and B is the set of members common to A and B. Symbolically, this is written $A \cap B$ and can for short be read "A cap B".

On the Venn diagram, Fig. 45, the shaded area represents the intersection of A and B.

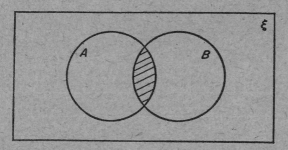

Fig. 45

If two sets in $\mathscr{E}$ have no common member, we have, Fig. 46, $A \cap B = \phi$ (an empty set) In such a case, the sets A and B are called *disjoint*.

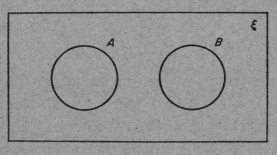

Fig. 46

Example. If $A = \{1, 2, 4, 5, 7, 8\}$, $B = \{1, 3, 5\}$ and $C = \{2, 4, 6, 8\}$, write down $A \cap B, A \cap C$ and $B \cap C$.

We have $A \cap B = \{1, 5\}$, for 1 and 5 are the only members common to set A and set C.

Similarly $A \cap C = \{2, 4, 8\}$

and $B \cap C = \phi$, for there are no common members of set B and set C.

Exercise 21

1. If $A = \{1, 3, 7, 9, 10\}$, $B = \{2, 3, 6, 7, 10\}$ and $C = \{1, 2, 3, 7\}$, find $A \cap B, A \cap C, B \cap C$.

2. Using the sets A, B, C in question 1 above, find $(A \cap B) \cap C$. [*Hint.* Work out $A \cap B = X$, say, firstly and then find $X \cap C$.]

3. Write down the following as single sets:

   (*a*) $\{1, 2, 4, 5, 7, 8\} \cap \{2, 3, 5, 7, 8, 9\}$,

   (*b*) {Prime numbers less than 10} $\cap$ {Odd integers},

   (*c*) $\{2, 4, 6, 7\} \cap \{3, 4, 5, 6\} \cap \{2, 3, 4, 5, 7\}$,

   (*d*) {Fish} $\cap$ {Porpoises}.

4. Draw two Venn diagrams with three sets A, B, C each, such that A and B are not disjoint and A and C are not disjoint. In one of the diagrams show B and C as disjoint; in the other diagram show B and C as not disjoint.

Is the statement "A and B are not disjoint" equivalent to the equation "$A \cap B \neq \phi$"? [*Note.* $\neq$ means "does not equal".]

5. A is the set of all rectangles and B is the set of all rhombuses. What is $A \cap B$?

6. All the boys in a form ride a bicycle. Does Sydney, who is a member of the form, ride a bicycle?

7. ABC is a triangle. P is the set of points equidistant from B and C; Q is the set of points equidistant from C and A. Illustrate the sets P, Q on a carefully drawn diagram. What is $P \cap Q$?

6. Union of Sets

In the section 5 above, we concentrated on studying the members which were common to two or more sets. We now turn our attention to consideration of *all* the members present in the sets under discussion.

Suppose we have the two sets $\{1, 2\}$ and $\{1, 3\}$. There are two elements (or members) in each set, but there are only three elements present, i.e. the numbers 1, 2, 3. The set $\{1, 2, 3\}$ is the *union* of the given sets.

If we invert the symbol for intersection, we have $\cup$, which stands for the union of two sets. Thus $A \cup B$ is read as "the union of A and B" or, more shortly, "A cup B".

Consider $A = \{1, 3, 5, 7\}$ and $B = \{1, 2, 3, 4\}$, then

$$A \cup B = \{1, 2, 3, 4, 5, 7\},$$

and this appears as the shaded area in Fig. 47.

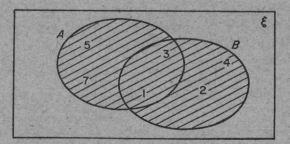

Fig. 47

It is important to note that members of the sets A and B only appear *once* in the union even if they appear in *both* of the sets. (In this example, the numbers 1 and 3 are in A and in B.)

Definition. The union of two sets is the total set of all elements which are present in the two given sets, each element being recorded only once.

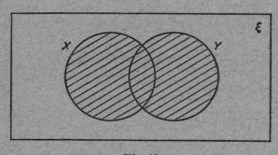

Fig. 48

In the Venn diagram, Fig. 48, the shaded area is $X \cup Y$.

We can extend the process of union and intersection, thus: $A \cup (B \cap C)$ means "the union of A with the set formed by the intersection of B and C". The part inside the brackets is evaluated first.

Example. $A = \{1, 3, 5, 8\}$, $B = \{2, 4, 6, 8\}$, $C = \{5, 6, 7, 8\}$. Find (a) $A \cup (B \cap C)$, (b) $(A \cup B) \cap C$, (c) $(A \cap C) \cup (B \cap C)$, (d) $(A \cap C) \cap (B \cap C)$.

(a) We work out $(B \cap C)$, the part in the brackets, first. We have

$$B \cap C = \{2, 4, 6, 8\} \cap \{5, 6, 7, 8\}$$

$$= \{6, 8\}$$

$$\therefore \quad A \cup (B \cap C) = \{1, 3, 5, 8\} \cup \{6, 8\}$$

$$= \{1, 3, 5, 6, 8\}$$

(b) Proceeding as in (a) above, we work out the inside of the brackets,
so
$$A \cup B = \{1, 3, 5, 8\} \cup \{2, 4, 6, 8\}$$
$$= \{1, 2, 3, 4, 5, 6, 8\}$$
$$\therefore \ (A \cup B) \cap C = \{1, 2, 3, 4, 5, 6, 8\} \cap \{5, 6, 7, 8\}$$
$$= \{5, 6, 8\}$$

There is an important point here. The results in (a) and (b) are not
identical, although the placings of the sets A, B, C and the cup and
cap symbols are the same. This means that the order of operation,
i.e. the positioning of brackets, is of fundamental significance.

(c) We have
$$A \cap C = \{1, 3, 5, 8\} \cap \{5, 6, 7, 8\}$$
$$= \{5, 8\}$$
and
$$B \cap C = \{2, 4, 6, 8\} \cap \{5, 6, 7, 8\}$$
$$= \{6, 8\}$$
$$\therefore \ (A \cap C) \cup (B \cap C) = \{5, 8\} \cup \{6, 8\}$$
$$= \{5, 6, 8\}$$
(d) Also $(A \cap C) \cap (B \cap C) = \{5, 8\} \cap \{6, 8\}$
$$= \{8\}.$$

Example. A club has 24 members, of whom 16 play tennis, 14 play
bridge and 9 play both games. How many do not play either game?

Here the simplest way is to draw a Venn diagram. We also
introduce a simple definition, namely, that $n(A)$ is the number of
elements in A.

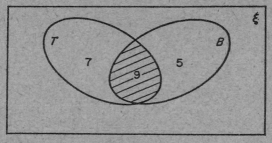

Fig. 49

Draw intersecting closed curves for $T=\{$tennis-players$\}$ and $B=\{$bridge-players$\}$. We know that they intersect because some members (actually 9) play both games. Put 9 in the intersection, (i.e. $n(T \cap B) = 9$) then the number of members of T not in $T \cap B$ is 7 and the number of members of B not in $T \cap B$ is 5.

$$\therefore \quad n(T \cup B) = 9+7+5 = 21$$

(this is the number of members in the union of T and B).

Now $\mathscr{E} = 24$ (the total number of members)

$$\therefore \quad n(\mathscr{E})-n(T \cup B) = 24-21 = 3.$$

This is the number who do not play either game.

Although in the above example we have for convenience used a minus sign on the last line of working, great discretion must be observed in this connection for reasons which will become apparent later. There is a better notation used later, in ch. 8 section 4 (The Complement of a Set).

Exercise 22

In questions 1–5 below, A represents the points within a square and B represents the points within a circle.

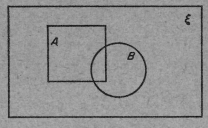

Fig. 50

1. Draw Fig. 50 twice. In the first case, shade $A \cap B$. In the second case, shade $A \cup B$.

2. Draw Fig. 51 and shade $A \cup B$.
   What is $A \cap B$?
   If we write $A \cap B = \{\ \}$, what must we put inside the braces?

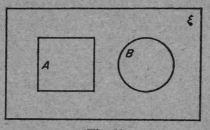

Fig. 51

3. Draw Fig. 52 twice and shade (i) $A \cap B$, (ii) $A \cup B$.

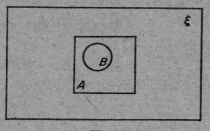

Fig. 52

4. In addition to A and B, defined above, we now also consider C, the set of points within a triangle. Draw diagrams, based on Fig. 53, to illustrate

        (a) $A \cap (B \cap C)$,   (c) $A \cup (B \cup C)$,
        (b) $A \cup (B \cap C)$,   (d) $A \cap (B \cup C)$.

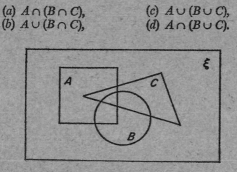

Fig. 53

[*Note*. Brackets are not necessary in (*a*) and (*c*) as there is no ambiguity as to the result. This will be shown later. In (*b*) and (*d*), brackets are essential.]

5. If $A = \{1, 2, 4, 6, 7, 9\}$, $B = \{1, 3, 6, 8, 9\}$, $C = \{2, 3, 4, 6, 8\}$, find $A \cap B$, $A \cap C$, $A \cup B$ and $A \cup C$.

Show, also, that

$$(a) \ (A \cap B) \cup (A \cap C) = A \cap (B \cup C)$$

$$(b) \ (A \cup B) \cap (A \cup C) = A \cup (B \cap C).$$

6. Using the sets A, B, C, defined in question 5 above, show that

$$A \cap (B \cap C) = (A \cap B) \cap C = (A \cap C) \cap B.$$

[This is true for all orders of A, B, C within the brackets and is true of *all* sets considered together. We can in fact write $A \cap B \cap C$.]

7. If A, B, C are any three sets shown as intersecting closed curves on a Venn diagram, show that $A \cup (B \cup C) = (A \cup C) \cup B$.

[Again, the positioning of brackets and the order of the letters is immaterial, and we can write $A \cup B \cup C$.]

8. Draw three intersecting sets P, Q, R on two Venn diagrams. On the first, show $P \cap (Q \cup R)$ and on the second, show $(P \cap Q) \cup R$ and hence demonstrate that they are not the same. [We certainly *cannot* omit brackets here!]

9. If $A \supset B$, demonstrate that $A \cap B = B$. What, in this case, is $A \cup B$, expressed as a single set?

10. If X, Y, Z are three sets, state which of the following statements are true and which are false:

$$(a) \ (X \cup Y) \supset Y, \quad (b) \ (X \cap Y) \supset Y, \quad (c) \ X \cap (Y \cup Z) \supset Y \cap Z.$$

Draw diagrams to illustrate each situation.

CHAPTER 7

INEQUALITIES

1. Inequalities

We introduce the symbols > meaning "greater than" and < meaning "less than". Two numbers a and b connected thus, $a > b$ or $b < a$, form an inequality. The relationships $a > b$ and $b < a$ are equivalent.

We know that 5 is greater than 2, i.e. $5 > 2$, and 2 is less than 5, i.e. $2 < 5$. On the number line we see at once that in the case of positive numbers the greater ones are to the right of the smaller ones. We extend the idea to cover *all* the numbers on the number line.

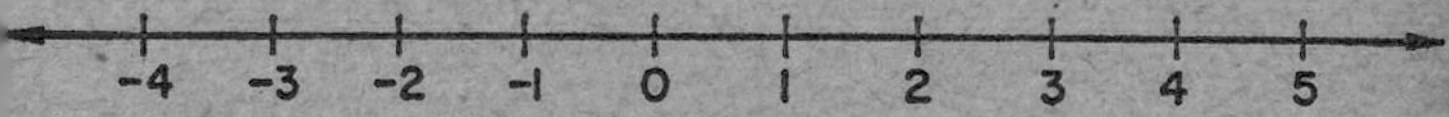

Fig. 54

e.g. $5 > 2$, $2 > -3$, $0 > -2$, $-1 > -4$.

The position is clarified by considering a man half-way up a flight of stairs on, say, a landing at zero level. Each step up is $+1$ and each step down is -1 (Fig. 55).

If he descends 1 step he is higher than if he descends 4 steps $\therefore$ $-1 > -4$.

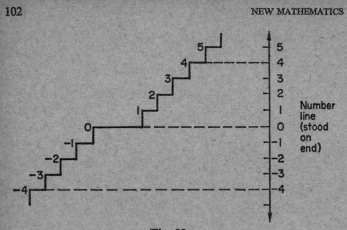

Fig. 55

Similarly a man who owes £100 is wealthier than a man who owes £400.

2. Set Notation for Inequalities

$\{x:x > 3\}$ means "the set of all numbers x such that x is greater than 3". On the number line this would appear:

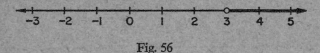

Fig. 56

Similarly $\{x:x < 2\}$ means "the set of all numbers x such that x is less than 2", and on the number line the situation is:

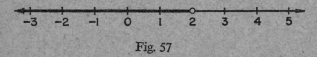

Fig. 57

The small open circle at the end of the set indicates that the end point on the number line is *not* a member of the set. If, however, we wish to include the end point we use modified symbols $\geqslant$ meaning "greater than or equal to" and $\leqslant$ meaning "less than or equal to". Thus $\{x : x \geqslant -1\}$ would appear on the number line as:

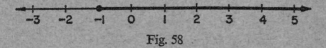

Fig. 58

This time the end point is included; the small solid circle indicates this.

Finally, we can have a situation such as $\{x : -1 < x \leqslant 2\}$,

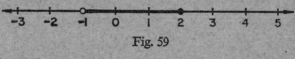

Fig. 59

or $\{x : x \geqslant 2 \text{ or } x \leqslant -2\}$.

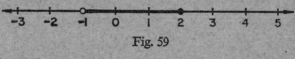

Fig. 60

The last two diagrams above are related to an important algebraic concept, quadratic inequalities.

Example. Consider $\qquad x^2 - 4 \geqslant 0$

We have $\qquad (x-2)(x+2) \geqslant 0$.

Now we know from elementary algebra $(+)(+) = (+)$ and $(-)(-) = (+)$

$$\therefore \quad x^2 - 4 \geqslant 0 \Leftrightarrow x + 2 \geqslant 0 \text{ and } x - 2 \geqslant 0$$

$$\textit{or } x + 2 \leqslant 0 \text{ and } x - 2 \leqslant 0.$$

The first pair is satisfied if $x \geqslant 2$, for in this case $x-2$ and $x+2$ are positive (or one is zero). The second pair is satisfied by $x \leqslant -2$, for in this case $x-2$ and $x+2$ are negative (or one is zero).

$$\therefore \quad \{x : x^2 - 4 \geqslant 0\} = \{x : x \geqslant 2\} \cup \{x : x \leqslant -2\},$$

for it is the union of the sets given by $x \geqslant 2$ and $x \leqslant -2$.

The number line picture is shown in Fig. 60 above.

Example. Consider $\{x:x^2-x-2<0\}$.

We have $(x-2)(x+1)<0$

$$x-2<0 \text{ and } x+1>0 \tag{1}$$

$$or \; x-2>0 \text{ and } x+1<0 \tag{2}$$

because $(-)(+)=(-)$

(1) implies $x<2$ and $x>-1$, which are possible at the same time,
(2) implies $x>2$ and $x<-1$, which are not possible at the same time.

$$\therefore \;\; \{x:x^2-x-2<0\} = \{x: \, -1<x<2\}.$$

3. Axioms of Inequalities

In order to be able to manipulate inequalities, we need four axioms, two for addition and two for multiplication. Those for addition are easily understood but those for multiplication require some care. The number line, however, is of considerable help in this connection.

Axiom 1. If the same number is added to each side of an inequality, the sides remain unequal in the same order. Thus if $a>b$, then $a+c>b+c$.

Axiom 2. If the same number is subtracted from each side of an inequality, the sides remain unequal in the same order. Thus if $a>b$, then $a-c>b-c$. This is fundamentally the same as axiom 1, for subtracting c is the same as adding $-c$. It was convenient here to state it separately.

Example.   Consider $4>-1$. Add 3 to each side, then

$$4+3>-1+3 \text{ (Axiom 1)}$$

i.e. $7>2$, which is clearly true.

On the number line, we have

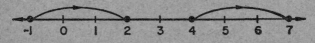

Fig. 61

Each point representing a side of the inequality is transposed three units to the right. Had we subtracted 3 from each side, the points would each have been moved three units to the left.

Axiom 3. If each side of an inequality is multiplied by the same *positive* number, the sign of inequality is unchanged.

Axiom 4. If each side of an inequality is multiplied by the same *negative* number, the sign of inequality is reversed. This axiom is important and is a source of difficulty to some students when they first meet it.

Example. Consider $3 > 2$. If we multiply each side by 2,

we have $\qquad 2.3 > 2.2$ (Axiom 3)

$\Rightarrow \qquad\qquad 6 > 4$, which is clearly true.

This is easily followed if we take the analogy of a man standing on the step 3 of Fig. 55 and a woman standing on step 2. If they double their step-height, the man will be on step 6 and the woman on step 4, so the man is still above the woman.

Now, however, consider $3 > 2$, with each side multiplied by -2. We have $\quad -2.3 < -2.2$ (Axiom 4: inequality sign reversed)

$\Rightarrow \qquad -6 < -4$

It is now of great help to look at Fig. 55. A man on step 3 who multiplies his level by -2 finds himself on step -6 whereas a woman on step 2 only finds herself on step -4, which is above step -6. The situation in each case is also easily seen on the number line:

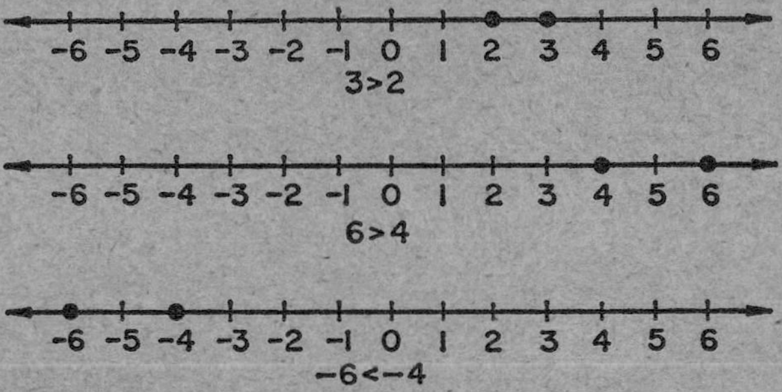

Fig. 62

Example. Solve $3x + 5 > 17$

We have $\quad\quad\quad 3x > 12$ $\quad\quad\quad\quad\quad$ (Subtracting 5: Axiom 2)

$\Rightarrow \quad\quad\quad\quad x > 4$ $\quad\quad\quad\quad\quad\quad\quad\quad\quad$ (Axiom 3)

Example. Solve $2 - 5x < 11$

We have $\quad\quad -5x < 9$ $\quad\quad\quad\quad\quad\quad\quad\quad\quad$ (Axiom 2)

$\Rightarrow \quad\quad\quad\quad 5x > -9$ $\quad\quad$ (Multiplying by -1: Axiom 4)

$\Rightarrow \quad\quad\quad\quad x > -1\cdot 8$

Example. Find all x such that $\{x : 2x - x^2 < 3x - 12\}$.

We have $\quad 2x - x^2 < 3x - 12$

$\Rightarrow \quad\quad -x^2 - x + 12 < 0$ $\quad\quad\quad$ (We omit the axioms once

$\Rightarrow \quad\quad x^2 + x - 12 > 0$ $\quad\quad\quad\quad$ they are fully understood.)

$\Rightarrow \quad\quad (x + 4)(x - 3) > 0$

$\Rightarrow \quad\quad x > 3 \text{ or } x < -4$

Exercise 23

Solve the inequalities and illustrate each solution on the number line (Nos. 1–5):

1. (*a*) $2x + 7 > 23$, $\quad\quad\quad\quad\quad$ (*b*) $S = \{x : 2x + 7 > 23\}$.
2. (*a*) $6x - 4 > x$, $\quad\quad\quad\quad\quad\quad$ (*b*) $S = \{x : 6x - 4 > x\}$.
3. (*a*) $2 - 3x \leqslant 2x + 3$, $\quad\quad\quad$ (*b*) $S = \{x : 2 - 3x \leqslant 2x + 3\}$.
4. $4x^2 - 9 < 0$,
5. $x^2 + 3x + 2 \leqslant 5(x + 2)$.
6. Illustrate S on the number line, where

$$S = \{x : x - 2 < 2\} \cap \{x : 3x > 5\}$$

[*Hint.* The first set is the such that $x < 4$ and the second is such that $x > 1\frac{2}{3}$. The sets intersect where $1\frac{2}{3} < x < 4$ and we illustrate as in Fig. 59, although, of course, the diagram is not the same.]

7. Illustrate S on the number line, where

$$S = \{x : 2x + 5 < 0\} \cup \{x : 3 - x < 0\}.$$

Show also on this line $S \cap T$, where

$$T = \{x : x^2 - 16 \leqslant 0\}.$$

Give the results (*a*) in the form of inequalities, (*b*) in set notation.

4. Graphical Representation of Inequalities involving two Variables (x, y).

Any straight line in Euclidean geometry is determined by two points on it. Furthermore, from elementary algebra, we know that any linear algebraic equation in x and y, such as $3x+2y = 5$, can be represented by a straight line graph, with axes $0x$, $0y$.

We now wish to determine the graphical representation of linear inequalities; for example, corresponding to $3x+2y = 5$ given above we might have any one of four inequalities

$$3x+2y > 5,\ 3x+2y \geqslant 5,\ 3x+2y < 5,\ 3x+2y \leqslant 5.$$

Let us first consider the graph of $3x+2y = 5$

x	0	1	2
y	2·5	1	−0·5

We take $x = 0, 1, 2$ in turn and table the corresponding values of y. We then draw the straight line through $(0, 2·5)$ and $(1, 1)$, the third point $(2, -0·5)$ being used only as a check. The line divides the plane into three sets of points, namely those on the line (S_1, say), those on one side of it (S_2, say) and those on the other side (S_3, say).

Now

$$S_1 = \{(x, y):3x+2y = 5\},$$

and this is read as "the set of ordered pairs such that $3x+2y = 5$". (For the definition of ordered pairs see Chapters 2 and 3.)

Suppose now we wish to graph

$$S_2 = \{(x, y):3x+2y > 5\}.$$

Here, if we put $x = 0$, we have $y > 2·5$, and if we put $x = 1$, we have $y > 1$. In each case the y-coordinate of $S_2 > y$-coordinate of S_1 and this is clearly true for all members of S_2. The graph of S_2 is the whole region of the x, y plane above $3x+2y = 5$. It is shaded in Fig. 63 and is called a half-plane. The line set S_1 is shown dotted as it is not part of the graph of S_2.

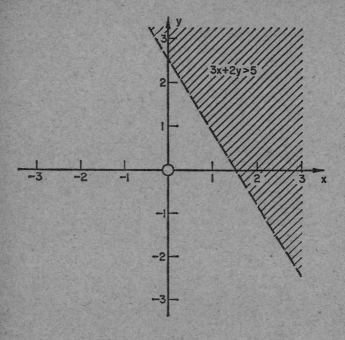

Fig. 63

Had we been asked to graph $3x+2y \geqslant 5$, the *line* set would have been continuous and not dotted for we would have needed

$$S_1 \cup S_2 = \{(x, y):3x+2y = 5\} \cup \{(x, y):3x+2y > 5\}.$$

The graphs of $3x+2y < 5$ and $3x+2y \leqslant 5$ are dealt with exactly similarly and would have led to the shading of the half-plane below S_1.

Two special cases need comment. Graphs such as $y \leqslant 2$ and $x > -3$, where only one of the variables is present, are easily understood. They are illustrated in Figs. 64A and 64B respectively.

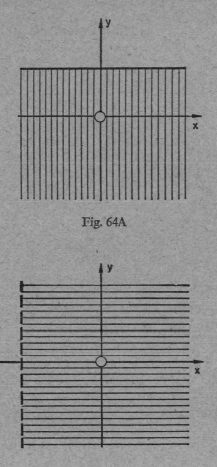

Fig. 64A

Fig. 64B

Suppose we now need a region enclosed by various conditions, e.g. all $(x, y) \in \{(x, y) : 3x + 2y \leqslant 5\} \cap \{(x, y) : x \geqslant -3\} \cap \{(x, y) : y \leqslant 2\}$. Using Figs. 63, 64A, 64B, suitably adapted, we have Fig. 65.

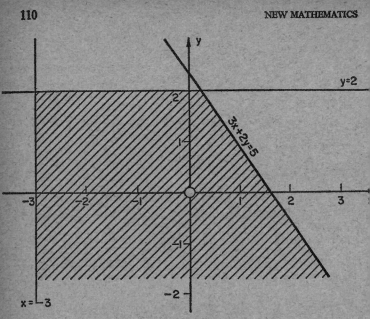

Fig. 65

It will be observed that the graph in this case is open below, but very often the region is entirely enclosed by a curve or a polygon. For example, $\{(x, y) : x^2 + y^2 \leqslant 4\}$ is the set of points entirely on or within the circle centred at the origin and having radius 2 units. It is easily seen (for those who do not know the coordinate geometry of the circle) that $x^2 + y^2 = 4$ is the circle referred to, by considering the right-angled triangle OPN below, Fig. 66A. P is any point (x, y) on the centre D, radius 2, and N is the foot of the perpendicular from P on Ox.

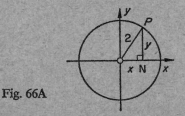

Fig. 66A

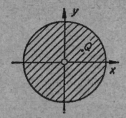

Fig. 66B

By Pythagoras's theorem,

$$ON^2 + NP^2 = OP^2 \Rightarrow x^2 + y^2 = 4.$$

Any point $Q(x, y)$ such that $OQ \leqslant OP$ lies within the circle or on it.

$$\therefore (x, y) \in \{(x, y): x^2 + y^2 \leqslant 4\}.$$

The whole set is mapped by the circumference and interior of the circle, Fig. 66B.

Exercise 24

Sketch the graphs of Nos. 1–3:

1. $y \geqslant x + 2$.
2. $2x + y < 0$.
3. $4y + 5 > 0$.

Illustrate the regions of Nos. 4–6:

4. $S = \{(x, y): x + y \geqslant 0\} \cap \{(x, y): y + 2 \geqslant 0\}$.
5. $T = \{(x, y): y \leqslant x + 1\} \cap \{(x, y): x \leqslant 4\} \cap \{(x, y): y > -1\}$.
6. $A = \{(x, y): (x^2 - 4)(y^2 - 9) < 0\}$.
7. Determine the region in the x, y plane such that

$$(x, y) \in \{(x, y): x^2 + y^2 \leqslant 9\} \cap \{(x, y): x^2 - 4 \leqslant 0\}.$$

5. Cartesian Product

Suppose we have two sets of integers,

$$A = \{1, 2, 3\} \text{ and } B = \{5, 7, 9, 11\}.$$

We define their Cartesian product $A \times B$ as the set of *ordered* pairs

$$(1, 5), (1, 7), (1, 9), (1, 11)$$
$$(2, 5), (2, 7), (2, 9), (2, 11)$$
$$(3, 5), (3, 7), (3, 9), (3, 11),$$

(where numbers in the first set must be put down first), in which every member of the first set is paired in turn with every member of the second set. The first set of elements (i.e. $\{1, 2, 3\}$) is called the *domain* and the second (i.e. $\{5, 7, 9, 11\}$) is called the *range*.

Similarly, if two young men $\{$Peter, George$\}$ take three young ladies $\{$Mary, Anne, Catherine$\}$ to a ball, the ways in which they can dance as mixed couples is

(Peter, Mary),   (Peter, Anne),   (Peter, Catherine),
(George, Mary),   (George, Anne),   (George, Catherine).

The general result for two sets

$$A = \{x_1, x_2, \ldots x_n\} \text{ and } B = \{y_1, y_2, \ldots y_m\}$$

is

$$(x_1, y_1), (x_1, y_2), \ldots (x_1, y_m)$$
$$(x_2, y_1), (x_2, y_2), \ldots (x_2, y_m)$$
$$(x_n, y_1), (x_n, y_2), \ldots (x_n, y_m)$$

There are n rows and m columns and thus there are nm ordered pairs.

Example. If $A = \{1, 2, 3\}$, mark the members of $A \times A$ on a graph. Indicate, also, the members (x, y) of $A \cap \{(x, y): y < x+1\}$.

Fig. 67A illustrates the nine points $(1, 1)$, $(1, 2)$, $(1, 3)$, $(2, 3)$ etc. and Fig. 67B shows those six which also satisfy the inequality $y < x+1$.

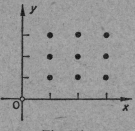

Fig. 67A

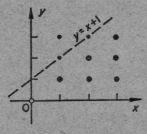

Fig. 67B

Example. Sketch, on the same diagram, the graphs of $x = 4$, $y = -1$, $x = y + 1$. Indicate on the diagram the region corresponding to the set S, where

$$S = \{(x, y) : x - y < 1\} \cap \{(x, y) : x < 4\} \cap \{(x, y) : y > -1\}.$$

If $(3 \cdot 5, \lambda) \in S$, find the *range* of values of λ. State the possible whole number values of λ and mark these corresponding points on the diagram.

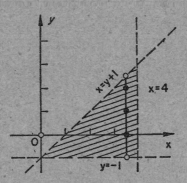

Fig. 68

The lower limit of $(3 \cdot 5, \lambda)$ is given by $x = 3 \cdot 5$, $y = -1$. The upper limit is given by $x = y + 1$, $x = 3 \cdot 5$, i.e. $y = 2 \cdot 5$. The required range is therefore

$$-1 < \lambda < 2 \cdot 5,$$

the end-points *not* being included in this case. The range is shown in Fig. 68. The only integral (whole number) values of λ are 0, 1, 2. (The end points give $\lambda = -1$ and $\lambda = 2 \cdot 5$; although one of these is integral it is inadmissible in our problem).

Exercise 25

1. If $A = \{1, 2, 3\}$ and $B = \{-2, 0, 2\}$, find all the members of $A \times B$ which lie within or on the circle $x^2 + y^2 = 9$.

2. Draw the graph of $4y = x^2$ for values of x such that $-3 \leqslant x \leqslant 3$. Shade the region $S = \{(x, y) : y \geqslant \frac{1}{4}x^2\} \cap \{(x, y) : y \leqslant 2\}$.

Given that $\mathscr{E}$ is the universal set of integers (including zero) and that S is defined in the equation above, mark on the graph all members of $(\mathscr{E} \times \mathscr{E}) \cap S$. Tabulate the Cartesian coordinates of these.

3. Draw the graph of $S = \{(x, y) : 2x + y \geqslant 3\} \cap \{(x, y) : x^2 + y^2 \leqslant 4\}$. If Z is the set of integers including zero, write down the Cartesian coordinates of the points (x, y) such that $x \in Z$, $y \in Z$, $(x, y) \in S$.

CHAPTER 8

SET THEORY

1. Standard Formulae

Before extending the work of the previous chapters, we list some important standard results.

$$A \cup A = A \qquad\qquad A \cap A = A$$
$$A \cup B = B \cup A \qquad\qquad A \cap B = B \cap A$$
$$A \cup \phi = A \qquad\qquad A \cap \phi = \phi$$
$$A \cup \mathscr{E} = \mathscr{E} \qquad\qquad A \cap \mathscr{E} = A$$
$$A \cup (B \cup C) = (A \cup B) \cup C \qquad A \cap (B \cap C) = (A \cap B) \cap C$$
$$A \cup (B \cap C) = (A \cup B) \cap (A \cup C) \qquad A \cap (B \cup C) = (A \cap B) \cup (A \cap C)$$

The first four pairs of formulae are almost self-evident. The other two pairs are generalizations of the results of Ex. 22, Qn. 5 on page 100.

Although it is not necessary to have recourse to Venn diagrams to demonstrate these formulae, the student is advised to draw some diagrams to illustrate several of the results.

We give below a mathematical proof of one of the formulae to illustrate the method used. It is straightforward but needs careful attention to detail.

Theorem. If A, B, C are three given sets, prove that
$$A \cap (B \cup C) = (A \cap B) \cup (A \cap C).$$

Proof. Let x be a member of a set.
Then $x \in A \cap (B \cup C) \Leftrightarrow (x \in A)$ and $(x \in B$ or C or both$)$

$$\Leftrightarrow (x \in A \text{ and } x \in B) \text{ or } (x \in A \text{ and } x \in C) \text{ or both}$$
$$\Leftrightarrow (x \in A \cap B) \text{ or } (x \in A \cap C) \text{ or both}$$
$$\Leftrightarrow x \in (A \cap B) \cup (A \cap C)$$
$$\therefore \quad A \cap (B \cup C) = (A \cap B) \cup (A \cap C).$$

The subtle point about this proof is the retention of the "or both" in three successive lines. There are rather neater ways of laying out the proof but they are not so easy to follow.

Similarly $A \cup (B \cap C) = (A \cup B) \cap (A \cup C)$.

If we consider $A \cap B \cap C$, we are in fact thinking of $\{x : x \in A, B \text{ and } C\}$ and it does not matter in what order we take A, B, C. It is not necessary, therefore, to write the brackets in $A \cap (B \cap C)$ or any other permutation of this. The Venn diagram (Fig. 69) makes this clear, the shaded area being $A \cap B \cap C$.

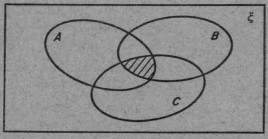

Fig. 69

2. The Intersection and the Union of three Sets

Likewise, if we consider $A \cup B \cup C$ we can omit brackets, such as occur in $A \cup (B \cup C)$, for $\{x : x \in A \text{ or } B \text{ or } C \text{ or any two or all three}\}$ and the order in which we write A, B, C is immaterial. The shaded area in Fig. 70 illustrates $A \cup B \cup C$.

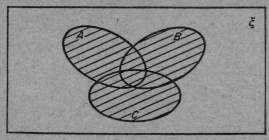

Fig. 70

It is easy to deduce two more simple results:

$$(A \cap B) \cap (A \cap C)' = A \cap B \cap C.$$
$$(A \cup B) \cup (A \cup C) = A \cup B \cup C.$$

We prove the first one.

Let x be a member of the set on the left hand side of the first equation.

Then $\qquad\qquad x \in A \cap B \Leftrightarrow x \in A$ and $x \in B$

but also $\qquad\qquad x \in A \cap C \Leftrightarrow x \in A$ and $x \in C$

$\therefore \quad x \in (A \cap B) \cap (A \cap C) \Leftrightarrow x \in A$ and $x \in B$ and $x \in C$

$$\Leftrightarrow x \in A \cap B \cap C.$$

The result then follows at once.

Exercise 26

1. Simplify (a) $A \cup (B \cap A)$, (b) $A \cap (B \cup A)$.

By methods similar to those employed in Section 1 above, prove that (Nos. 2–4):

2. $A \cap B = B \cap A$.

3. $(A \cup B) \cup (A \cup C) = A \cup B \cup C$.

4. $(A \cup B) \cap (A \cup C) = A \cup (B \cap C)$.

5. Simplify (a) $[(A \cap B) \cup (A \cap C)] \cap [A \cup B]$,

$\qquad\qquad$ (b) $[(A \cup B) \cap (A \cup C)] \cup [(A \cap B) \cup (A \cap C)]$.

[*Hint.* Sketches will help.]

3. Highest Common Factor and Lowest Common Multiple

Suppose we wish to find the H.C.F. and L.C.M. of the numbers 210; 595; 770. Let the sets of their prime factors be A, B, C respectively. We have

$$210 = 2.3.5.7; \quad 595 = 5.7.17; \quad 770 = 2.5.7.11$$

$$\therefore \quad A = \{2, 3, 5, 7\}; \quad B = \{5, 7, 17\}; \quad C = \{2, 5, 7, 11\}.$$

The factors of the H.C.F. are the members of $A \cap B \cap C$ for these are the only members common to all three of A, B, C. But

$$A \cap B \cap C = \{5, 7\}$$

$$\therefore \quad \text{H.C.F.} = 5.7 = 35.$$

Similarly $A \cup B \cup C$ gives the factors of the L.C.M. But

$$A \cup B \cup C = \{2, 3, 5, 7, 11, 17\}$$

$$\therefore \quad \text{L.C.M.} = 39270.$$

There is a difficulty when there are repeated factors of numbers for which we are finding the H.C.F. and L.C.M., for we must recall the necessity to be able, in mathematics, to distinguish between members of a set (page 87). We overcome this snag by introducing suffixes.

Consider the numbers 24; 28; 45. We have[13]

$$24 = 1.2_a.2_b.2_c.3; \quad 28 = 1.2_a.2_b.7; \quad 45 = 1.3_a.3_b.5$$

Using the notation above

$$A \cap B \cap C = 1,$$

i.e. A, B, C have no common factor (other than unity).

$$\therefore \quad \text{H.C.F.} = 1$$

$$A \cup B \cup C = \{2_a, 2_b, 2_c, 3_a, 3_b, 5, 7\}$$

$$\therefore \quad \text{L.C.M.} = 2^3.3^2.5.7 = 2520$$

Example. Find the H.C.F. and L.C.M. of $6a^2b$, $3a^3c$, $8abc^2$. We have

$$A = \{2_1, 3, a_1, a_2, b\}$$

$$B = \{3, a_1, a_2, a_3, c_1\}$$

$$C = \{2_1, 2_2, 2_3, a_1, b, c_1, c_2\}$$

$$\therefore \quad A \cap B \cap C = \{a_1\}$$

i.e.   H.C.F. $= a$

Also

$$A \cup B \cup C = \{2_1, 2_2, 2_3, 3, a_1, a_2, a_3, b, c_1, c_2\}$$

i.e.   L.C.M. $= 2^3.3.a^3.b.c^2 = 24a^3bc^2.$

Clearly, in an example of this kind, the solution by ordinary algebra is shorter, but the above method brings out the logical reasoning underlying set theory.

---

[13] We put in the factor 1 only when there is no other common factor.

Exercise 27

Find, using methods similar to those above, the H.C.F. and L.C.M. of

1. (a) 1155;   462;   560,   (b) 42;   48;   72;   102.
2. (a) x^2yz, y^3z^2, x^3z, ---   (b) $6a^3bc$, $16ac^2$, $30b^2c^2$.
3. $4a^2 - 4ab$, $6ab - 6b^2$.   [*Hint.* Factorize the expression.]
4. 512;   768;   1536.

4. The Complement of a Set

If $\mathscr{E}$ is the universal set and if A is a set in $\mathscr{E}$, then the complement of A is called A' and is given by

$$A \cup A' = \mathscr{E}.$$

Thus A' contains all the members of $\mathscr{E}$ which are not in A, and *vice versa*.

It follows that $x \in A \Rightarrow x \notin A'$, for A, A' have no common members. On the Venn diagram, Fig. 71, the situation is easily understood.

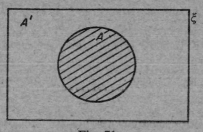

Fig. 71

Hence $A \cap A' = \phi$ and $A \cup A' = \mathscr{E}$.

Theorems.   (i) $A' \cap B' = (A \cup B)'$;   (ii) $A' \cup B' = (A \cap B)'$, where $(A \cup B)'$ and $(A \cap B)'$ are the complements of $A \cup B$ and $A \cap B$ respectively.

We shall give an intuitive proof of (i) using Venn diagrams and then prove it formally. Theorem (ii) can be proved similarly and is given in Exercise 28 for the reader. These two theorems are called de Morgan's laws.

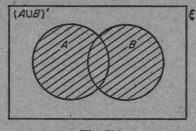

Fig. 72A

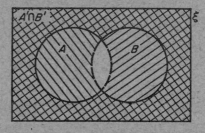

Fig. 72B

In Fig. 72A, $A \cup B$ is shaded and hence $(A \cup B)'$ is the unshaded area. In Fig. 72B, the cross-hatched area is $A' \cap B'$ (for the figure has two overlapping holes A and B cut out of it; the complement of A, i.e. A', is shaded top right to bottom left; the complement of B, i.e. B', is shaded top left to bottom right; their intersection $A' \cap B'$ is therefore the cross-hatched region of $\mathscr{E}$). Now the *unshaded* area of Fig. 72A is the same as the *cross-hatched* area of Fig. 72B.

$$\therefore \quad (A \cup B)' = A' \cap B'.$$

There is no need for A and B to overlap (i.e. they need have no common members) but in this case the demonstration is almost trivial.

We now give a theoretical proof.

Let x be a member of a set, then

$$x \in (A' \cap B') \Leftrightarrow (x \in A') \text{ and } (x \in B')$$
$$\Leftrightarrow (x \notin A) \text{ and } (x \notin B)$$
$$\Leftrightarrow x \notin (A \text{ or } B \text{ or both})$$
$$\Leftrightarrow x \notin (A \cup B)$$
$$\Leftrightarrow x \in (A \cup B)'.$$

Hence $\qquad A' \cap B' = (A \cup B)'.$

We conclude this section with rather a harder problem, demonstrating with Venn diagrams.

Theorem. $(A \cap B)' = (A' \cap B') \cup (A \cap B') \cup (A' \cap B).$

Consider the expression on the right hand side. We know from above that $A' \cap B'$ is the *cross-hatched* area of Fig. 72B. Now $A \cap B'$ and $A' \cap B$ are the *shaded areas* in Fig. 73A and Fig. 73B respectively. (The shading accords with Fig. 72B.)

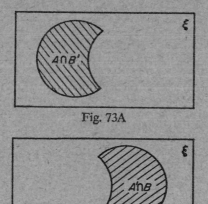

Fig. 73A

Fig. 73B

Hence by addition the union of $A' \cap B'$, $A \cap B'$, $A' \cap B$ is the shaded area of Fig. 74A.

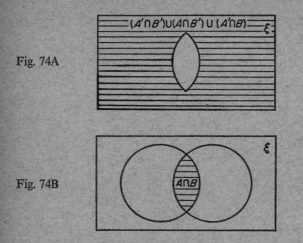

Fig. 74A

Fig. 74B

We see at once that this is the complement of $A \cap B$, the shaded area in Fig. 74B. Hence the theorem has been demonstrated to be true.

The methods used in this chapter are based on the work (published in 1854) of George Boole (*d.* 1864), an English mathematician, and the subject is called Boolean algebra. Although neglected for more than a century, the development of electronic computers has led to a sudden revival of interest in his studies for they are well suited to the needs of these machines.

Exercise 28

1. *A* and *B* are the sets of points with intersecting circles, Fig. 75;

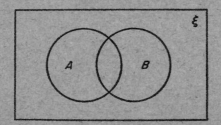

Fig. 75

and $\mathscr{E}$ is the universal set. Draw diagrams and shade them to illustrate (a) $A \cap B'$, (b) $A' \cup B$, (c) $A' \cap B'$, (d) $A' \cup B'$, (e) $(A \cup B)'$.

2. A and B are the sets of points within the triangle and circle respectively, in Fig. 76, and $\mathscr{E}$ is the universal set. Draw separate diagrams and shade them suitably to illustrate
(a) $A \cup B'$, (b) $(A \cup B)'$, (c) $A \cap B'$, (d) $(A \cap B)'$, (e) $(A' \cap B)$.

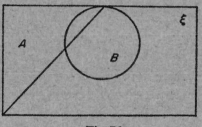

Fig. 76

3. If $\mathscr{E} = \{1, 2, 3, 4, 5, 6, 7, 8\}$, $A = \{1, 3, 5, 7\}$, $B = \{2, 3, 4, 6, 7\}$, write down (a) A', (b) B', (c) $A' \cap B'$, (d) $A' \cup B'$, (e) $A \cap B'$, (f) $(A \cup B)'$, (g) $(A \cap B)'$.

4. Prove that $A' \cup B' = (A \cap B)'$, (i) by Venn diagrams, (ii) theoretically.

5. Show,[14] using Venn diagrams, that

$$A \cup (B \cap C) = (A \cup B) \cap (A \cup C).$$

6. $\mathscr{E}$ is the universal set $\{a, b, c, d, e, f\}$ and $A = \{a, b, c, d\}$, $B = \{c, d, e, f\}$, $C = \{a, b, e, f\}$ are subsets of $\mathscr{E}$. Find the set $(A' \cap B) \cup (A \cap B')$. What do you observe?

7. A survey of 119 people taking a train to London elicited the following useless information. 60 were going to a football match, 55 were going to an evening show and 90 were going out to dinner. Of these, 48 were going to a show and to dinner, 37 were going to a match and to dinner. Of these subsidiary groups, 10 were going to

[14] A *theoretical* proof was asked for in Exercise 26.

all three affairs. If 4 of the people questioned were not doing any of these things, find how many were going

(a) to the football match only,
(b) to the football match and to a show but not to dinner.

The question is solved in outline below.
Let

D = {diners}, F = {football fans}, S = {people going to a show}.

The universal set $\mathscr{E}$ has 119 members. Draw a Venn diagram, Fig. 77.

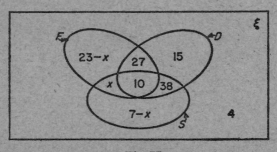

Fig. 77

We have $n(F \cap D \cap S) = 10$, and this is entered first. Also, $n(F \cap D) = 37$, so the number in $F \cap D$ but not in $F \cap D \cap S$ is $37 - 10 = 27$. We enter this next. Proceeding similarly, we have $n(D \cap S) = 48$ and $n(F \cap D \cap S) = 10$; hence the number in $n(F \cap D) - n(F \cap D \cap S) = 48 - 10 = 38$. This also is entered on the diagram. We can now complete D, for the remainder dining is $90 - 27 - 38 - 10 = 15$. The difficulty in the problem is now apparent. We do not know $F \cap S$. We put x in the part of $F \cap S$ which is not in $F \cap D \cap S$, i.e. $x = n(F \cap S) - n(F \cap D \cap S)$.

Then
$$n(F) = 60 - 27 - x - 10 = 23 - x$$
and
$$n(S) = 55 - 38 - x - 10 = 7 - x.$$

These also are put on the Venn diagram. All we need to do now is to add the elements in Fig. 77 and to equate the result to the number of people concerned ($119 - 4 = 115$). This gives us x. The reader is left to show that $x = 5$ (the number going to the match *and* to a show but *not* to a dinner). What is the number going *only* to the match?

8. In a certain school, 23 candidates were taking G.C.E. Advanced Level English Literature and 29 were taking History. Some were taking Geography; 10 candidates were doing English Literature only, 15 were doing History only, 6 were doing English Literature *and* History only, 4 were doing Geography *and* English Literature only and 10 were doing Geography only. (*a*) How many candidates altogether were taking Geography? (*b*) How many candidates were taking all three subjects? (*c*) How many candidates altogether were involved in the G.C.E. examinations?

5. Symmetric Difference

We define the symmetric difference of two sets A and B as the union of the set of elements in A but not in B together with those in B but not in A. Symbolically, this is written $A \triangle B$.

On the Venn diagram, Fig. 78, the region is shaded.

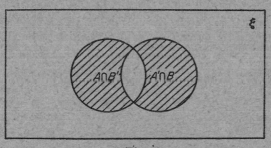

Fig. 78

The set of elements in A but not in B is $A \cap B'$, and the set in B but not in A is $A' \cap B$.

$$\therefore \quad A \triangle B = (A \cap B') \cup (A' \cap B).$$

This is represented by the shaded area in Fig. 78. Although it is possible to use the set notation already explained, it is easier to manipulate the symmetric difference formula after mastering the abbreviated notation given in Section 6 below.

6. Abbreviated Notation

It will have become apparent that the notation of Boolean algebra is rather tiresome to write. We accordingly introduce a widely-used symbolism which appears similar to ordinary algebra. It is, however, subject to rules which differ to some extent from those of the latter. The new notation was not mentioned earlier as it was necessary for the reader to be conversant with set theory before taking any short cuts.

We replace $A \cap B$ by AB and $A \cup B$ by $A+B$. We also substitute 1 for $\mathscr{E}$ (the universal set) and 0 for ϕ (the null set). Our standard results are modified thus:

	Set Notation	Abbreviated Notation	
1.	$A \cup A = A$	$A+A = A$	
2.	$A \cup B = B \cup A$	$A+B = B+A$	*
3.	$A \cup (B \cup C) = (A \cup B) \cup C$	$A+(B+C) = (A+B)+C$	*
4.	$A \cap A = A$	$AA = A$	
5.	$A \cap B = B \cap A$	$AB = BA$	*
6.	$A \cap (B \cap C) = (A \cap B) \cap C$	$A(BC) = (AB)C$	*
7.	$A \cup \phi = A$	$A+0 = A$	*
8.	$A \cap \phi = \phi$	$A.0 = 0$	*
9.	$A \cap (B \cup C) = (A \cap B) \cup (A \cap C)$	$A(B+C) = AB+AC$	*
10.	$A \cup (B \cap C) = (A \cup B) \cap (A \cup C)$	$A+BC = (A+B)(A+C)$	
11.	$A \cup \mathscr{E} = \mathscr{E}$	$A+1 = 1$	
12.	$A \cap \mathscr{E} = A$	$A.1 = A$	*
13.	$A \cup A' = \mathscr{E}$	$A+A' = 1$	
14.	$A \cap A' = \phi$	$AA' = 0$	

The eight results, each of which bears an asterisk alongside, are similar to those of ordinary algebra. The six results which are not so indicated require care but are easily understood provided one visualizes the corresponding Venn diagrams. Of these fourteen results, the hardest to visualize is No. 10. The reader is advised to memorize all of these as rapidly as possible.

Example. Prove that, in Boolean algebra,

$$(A+B)(B+C)(C+A) = BC+CA+AB.$$

We have

$(A+B)(B+C)(C+A)$	$= [(A+B)(A+C)](B+C)$	Results 2 and 5
	$= (A+BC)(B+C)$	Result 10
	$= AB+AC+B^2C+BC^2$	Results 9 and 5
	$= AB+AC+BC+BC$	Result 4
	$= AB+BC+CA$	Results 1 and 2

(as $BC+BC = BC$, for all the elements of BC are merely repeated).
Once familiarity with procedure has been gained, it is not necessary to quote reasons.

Exercise 29

Simplify the expressions in questions 1–14:

1. $A+(A+B)$.
2. $1-A$.
3. $0(A+B)$.
4. $1+A$.
5. $B(1+A)$.
6. $A+AB$. [*Hint.* This is $A(1+B)$.]
7. $A(A+B)$.
8. $(A+B)(A+B)$.
9. $(A+B)(B+C)$.
10. $(A+B)'(A+B)$.
11. $A(A+B)(B+C)$.
12. $A'(A+B)$.
13. $(A+B')(A'+B)$.
14. $(A+B)^2(A+C)$.

15. Where, in the earlier set notation, has the theorem
    $A'+B' = (AB)'$ been discussed?
Complete the corresponding theorem $A'B' = (\quad)'$.
16. Prove that $A'B'+AB'+A'B = (AB)'$.
[This has been shown earlier, using Venn diagrams (*where*?) but it is quite easily proved using the standard results, together with the theorem $A'+B' = (AB)'$ (question 15 immediately above).

We start with $A'B'+AB'+A'B$	$= A'(B+B')+AB'$	Results 2, 9
	$= A'+AB'$	Result 13
	$= (A'+A)(A'+B')$	Result 10

The reader is left to complete the question, which is now almost self-evident.]

17. Prove, without the use of Venn diagrams, and in the abbreviated notation shown above, that

$$A'B' = (A+B)'.$$

[This is the second theorem asked for in question 15 above.]
18. Prove that $(A+B')(B+C')(C+A') = ABC+A'B'C'$.
Note. The reader is advised to use this notation with extreme care if there is likelihood of ordinary algebra appearing in the same working.

7. Manipulation with the Symmetric Difference Formula

We have already seen that the symmetric difference formula is, in set notation,

$$A \triangle B = (A \cap B') \cup (A' \cap B).$$

In the abbreviated form, this becomes

$$A \triangle B = AB' + A'B.$$

The formulae given in Ex. 29, Nos. 15 and 17, namely

$$A' + B' = (AB)'$$

and $$A'B' = (A+B)'$$

are worth memorising. They are used here and later in the book.

Theorem. $(A \cap B) \triangle (A \cap C) = A \cap (B \triangle C)$.
  In abbreviated form, this is

$$(AB) \triangle (AC) = A(B \triangle C).$$

We have

$$
\begin{aligned}
(AB) \triangle (AC) &= (AB)(AC)' + (AC)(AB)' \\
&= AB(A'+C') + AC(A'+B') \quad \text{from above} \\
&= AA'B + ABC' + AA'C + AB'C \\
&= ABC' + AB'C \quad\quad\quad\quad \text{for } AA' = 0 \\
&= A(BC' + B'C) \\
&= A(B \triangle C).
\end{aligned}
$$

Exercise 30

1. By considering $\mathscr{E} = \{a, b, c, d, e\}$, $A = \{a, b, c\}$, $B = \{c, d, e\}$ verify that in this case, $A \triangle B = (A \cap B') \cup (A' \cap B)$.

2. If $A = \{1, 3, 4, 6, 9\}$ and $B = \{2, 3, 4, 5, 9\}$, write down the value of $A \triangle B$, given that the universal set consists of the natural numbers less than 10.

Check that, in this case,

$$(AB) \triangle (AC) = A(B \triangle C).$$

3. Prove that

$$(A \triangle B)(A \triangle C) = AB'C' + A'BC.$$

4. Prove that

$$(A+B) \triangle (A+C) = A'(B \triangle C).$$

Deduce that

$$(AB) \triangle (AC) + (A+B) \triangle (A+C) = B \triangle C.$$

5. Prove that

$$(A \triangle B)' = AB + A'B' = A' \triangle B = A \triangle B'.$$

6. Simplify

$$(A \triangle B) \triangle (A \triangle C).$$

CHAPTER 9

MATHEMATICAL SYMBOLIC LOGIC

1. Logic

Even the most experienced authors run out of ideas sometimes and, having planned this chapter and feeling that the first section was of necessity somewhat tedious, the present writer started to ferret around for simple ideas. His friend, John Ward, gave the following constructive example and the author's son, Peter, promptly produced another which has a curious quirk in it.

Consider the assertions

    a All Christians are good people,

    b All Buddhists are good people,

∴ *c* All Christians are Buddhists.

We have endeavoured to deduce a statement from two given statements, but there is a flaw in our logic. It is easily seen in the Venn diagrams, Figs. 79A and 79B.

Let $\mathscr{E} = \{$good people$\}$, $C = \{$Christians$\}$, $B = \{$Buddhists$\}$.

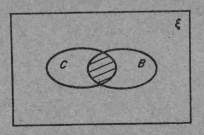

Fig. 79A

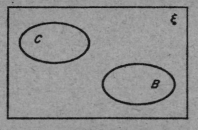

Fig. 79B

Clearly in both figures, the condition is met that the sets B, C lie entirely in the universal set $\mathscr{E}$, but whereas in Fig. 79A it is *possible* that there may be *some* Christians who are Buddhists, there is no *necessity* for this at all, for the situation in Fig. 79B, where $C \cap B = \phi$, is equally valid as a deduction. In fact, it would be reasonable to say that to all intents and purposes the latter figure represents the true life state of affairs.

Consider now the following assertions

   a Coal burns in the hearth,

   b Wood burns in the hearth,

∴ *c* Coal is wood.

Nonsense, we first say; but is it? Coal is, after all, wood which has been buried under intense pressure for millions of years. The logic is, of course, as specious as in the earlier example. It just so happens that the correct Venn diagram *could* be drawn as in Fig. 80.

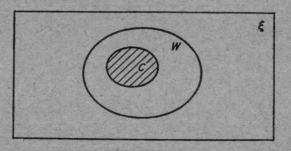

Fig. 80

Let $\mathscr{E} = \{$material burned in a hearth$\}$, $W = \{$wood$\}$, $C = \{$coal$\}$, then we see that we can say $W \supset C$.

The theory of mathematical symbolic logic is intimately related to the theory of Boolean algebra, explained in the two preceding chapters. It is curious that investigation of the subject should only have occurred systematically as late as the XIXth century, for we constantly use logical reasoning in the problems of everyday life—and we have been doing this for thousands of years. A non-mathematical approach to logic has, of course, existed for many centuries but we have proved to be markedly tardy in developing a mathematical shorthand for the subject.

We shall see that the affinity between set theory and symbolic logic is very close indeed. So intimately linked are the ideas, it would be true to say that they are slightly different approaches to the same subject, set theory arising from considerations of objects, and logic from considerations of statements.

We continue our investigations with another example:

a $PQRS$ is a rhombus.

b   The opposite angles of the quadrilateral $PQRS$ are equal.

Here $a \Rightarrow b$

is true, for we know that the opposite angles of a rhombus are equal. It does *not* follow, however, that the converse

$$b \Rightarrow a$$

is true; for, if statement b is the original assertion (i.e. if the opposite angles of $PQRS$ are equal), then $PQRS$ is a parallelogram (which may or may not be a rhombus) and statement a is not valid. It *is* true that

$$b' \Rightarrow a',$$

i.e. that if the opposite angles of quadrilateral $PQRS$ are not equal, then $PQRS$ is not a rhombus.

In this example, a' is said to be the negation of a and b' is the negation of b. (These correspond to the complements A', B' of sets A, B).

We can present this statement equally easily in set notation. Suppose

$A = \{\text{rhombuses}\}$, $B = \{\text{quadrilaterals with opposite angles equal}\}$:

then $A \subset B$ gives rise to $B' \subset A'$.

These results are true in general.

Theorem. If we have two statements a, b such that $a \Rightarrow b$, then $b' \Rightarrow a'$. This can be written

$$(a \Rightarrow b) \Leftrightarrow (b' \Rightarrow a').$$

The corresponding result for two sets A, B is

$$A \supset B \Leftrightarrow B' \supset A'.$$

Alternatively $$A \subset B \Leftrightarrow B' \subset A'.$$

The truth of this is easily demonstrated on a Venn diagram and verification is left as an exercise for the reader. (See Exercise 31.)

2. Notation

We now define and extend our logical symbols.

If a is an assertion, then a' is the negation of a (i.e. the statement denying the truth of a).

If a and b are assertions, then

(i) $a \wedge b$ is the statement asserting *both* a and b,

(ii) $a \vee b$ is the statement asserting *either* a or b *or both* a and b.

The "a and b" (i.e. $a \wedge b$) corresponds exactly with $A \cap B$ in set theory, as can be seen at once by considering Fig. 45. The "a or b" (i.e. $a \vee b$) corresponds exactly with $A \cup B$ in set theory, as can be seen from Fig. 48. Great care is needed in using the *or* symbol, $\vee$, for we assert that this means "*either* one statement *or* the other statement *or* both statements".

We see at once that the algebra of sets lends itself to symbolic logic. The symbols $\wedge$ and $\cap$ look very similar and so do $\vee$ and $\cup$; from the point of view of remembering them, this is convenient.

Before proceeding further we shall examine some examples to find whether conclusions can be drawn logically from assertions.

Example. Suppose   *a.* All intelligent people eat apples.
                    b. Some boys eat apples.
                    c. Some boys are intelligent.
Is it true that $a \wedge b \Rightarrow c$?

We can see the situation in a Venn diagram.

$A = \{$apple eaters$\}$, $P = \{$intelligent people$\}$, $B = \{$boys$\}$.

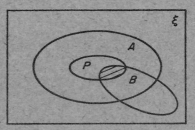

Fig. 81A

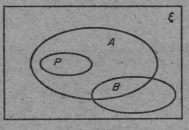

Fig. 81B

We have $a \Rightarrow A \supset P$, but
           $b \Rightarrow$ some boys (not necessarily all) are in A.
It follows that the category we want for c (i.e. $P \cap B$) may have some
members, Fig. 81A, or may be a null set, Fig. 81B. We cannot say,
therefore, that $a \wedge b \Rightarrow c$.

If statement a had been transposed we would have been able to
draw such a conclusion. We reset our problem and try again.

                    a. All apple eaters are intelligent.
                    b. Some boys eat apples.
                    c. Some boys are intelligent.

It is now true that $a \wedge b \Rightarrow c$.

The Venn diagram, Fig. 82, illustrates this. The set P of intelligent people now includes all apple eaters (we ignore insects and birds) and the set we need is $P \cap B$ (shown shaded).

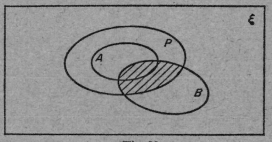

Fig. 82

As some boys are in set A (apple eaters), which is entirely in set P, it follows that some boys are in set P.

Exercise 31

State whether the conclusions in Nos. 1–7 are logical. Draw a Venn diagram for each question, to illustrate the situation. The fact that you may disagree with the sentiments expressed in the initial statements does not affect the logic.

1. An isosceles triangle can be drawn with its vertical angle of any size between 0° and 180°.
   Some triangles are right-angled.
   Therefore some right-angled triangles are isosceles.
2. All people like to have more than other people.
   Socialists do not like anyone to have more than anyone else.
   Therefore socialists do not exist.
   (The question is not intended to be taken too seriously!)
3. Some programmes on television lead viewers to make things.
   Television makes viewers lazy.
   Therefore television causes lazy viewers to make things.
4. French people speak French.
   German people speak German.
   Therefore Italian people speak Italian.

5. Some numbers are multiples of 2.
   Some numbers are multiples of 3.
   Some numbers are multiples of 4.
   Therefore some numbers are multiples of 24.
6. All businessmen go into Manchester by car.
   Cars may not be parked in Manchester.
   Therefore all businessmen go into Manchester by public transport.
7. Explorers get surrounded by flies.
   There are no flies on a clever man.
   Therefore a clever man never explores.

In Nos. 8–10, assuming that the initial statements are true, write down a logical conclusion in each case.

8. All boys have pocket-money.
   Some boys smoke cigarettes.
9. All women like new hats.
   Some women do not wear hats.
10. N is an even number divisible by 3 and by 7.
11. Travellers go up North.
    Travellers go down South.
    Travellers go up to town.
    Therefore travellers from Liverpool go down to London—or is it up?
12. Show, with the aid of Venn diagrams, that

$$A \subset B \Leftrightarrow B' \subset A'.$$

3. Truth Tables

We now consider propositions which are either true or false. If the statement a is true, it is said to have a *truth value* 1, and if it is false it has a truth value 0. Hence, if $a = 1$, then $a' = 0$ and if $a = 0$, then $a' = 1$. These can be written

$$a = 1 \Leftrightarrow a' = 0 \text{ and } a = 0 \Leftrightarrow a' = 1.$$

Extensions are equally valid, of course, e.g.

$$(a \wedge b) = 0 \Leftrightarrow (a \wedge b)' = 1 \text{ and } (a \vee b) = 0 \Leftrightarrow (a \vee b)' = 1.$$

The idea can obviously be related to set theory, for truth value 1 corresponds to set membership and truth value 0 corresponds to non-membership.

By this time it must have become abundantly clear to the reader that propositions need not be true to life. Logic is merely concerned with correctly deduced conclusions being drawn from given assertions. If the initial assertions are eccentric, we can hardly expect the results to be anything else!

Before we proceed to draw up truth tables, we must be very sure that we understand the meaning of $a \wedge b$ (a and b) and ($a \vee b$ (a or b or both a and b).

(i) $a \wedge b$. If $a = 1$ and $b = 1$, then $a \wedge b = 1$, for a and b are both 1. For all other cases, i.e.

$$\left. \begin{array}{l} a = 1, b = 0 \\ a = 0, b = 1 \\ a = 0, b = 0 \end{array} \right\}, \text{ we have } a \wedge b = 0,$$

for a and b are not *both* 1. The situation corresponds exactly to the shaded area in Fig. 83.

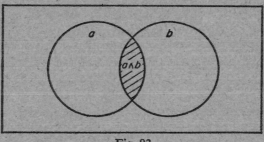

Fig. 83

(ii) $a \vee b$. This is more subtle. We recall that $a \vee b$ means that success is achieved if *either* $a = 1$, or $b = 1$, or a and b are both 1.

Thus, if

$$\left. \begin{array}{l} a = 1, b = 1 \\ a = 1, b = 0 \\ a = 0, b = 1 \end{array} \right\}, \text{ we have } a \vee b = 1.$$

The only failure is $a = 0$, $b = 0$, which gives $a \vee b = 0$. The situation corresponds exactly to the shaded area in Fig. 84.

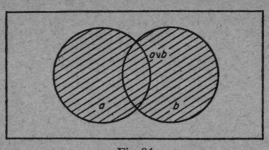

Fig. 84

Suppose we have two assertions, a and b, then a can be 1 or 0 and b can be 1 or 0. Defining "and" and "or" as in (i) and (ii) on page 133, we have all possible combinations of $a \wedge b$, $a \vee b$, in the table:

a	b	$a \wedge b$	$a \vee b$
1	1	1	1
1	0	0	1
0	1	0	1
0	0	0	0

and these are called the truth tables for $a \wedge b$ and $a \vee b$.

Proceeding further, we have

a	b	$a \vee b$	$(a \vee b)'$	a'	b'	$a' \wedge b'$
1	1	1	0	0	0	0
1	0	1	0	0	1	0
0	1	1	0	1	0	0
0	0	0	1	1	1	1

We observe that columns 4 and 7 in this table are the same, but we have covered every possible combination of a and b, therefore we have proved that
$$(a \vee b)' \Leftrightarrow a' \wedge b'.$$

It can similarly be proved that

$$(a \wedge b)' \Leftrightarrow a' \vee b'$$

and this is left as an exercise to the reader (see p. 121).

Theorem. If $a \Rightarrow b \vee c$, then $b' \wedge c' \Rightarrow a'$.

Proof: We have already seen that

$$[a \Rightarrow b] \Leftrightarrow [b' \Rightarrow a']$$

$$\therefore \quad [a \Rightarrow b \vee c] \Leftrightarrow [(b \vee c)' \Rightarrow a']$$

$$\Leftrightarrow [b' \wedge c' \Rightarrow a'], \text{ from above.}$$

4. Standard Results

	and		*or*
1A	$a \wedge a \Leftrightarrow a$	1B	$a \vee a \Leftrightarrow a$
2A	$a \wedge a' \Leftrightarrow 0$	2B	$a \vee a' \Leftrightarrow 1$
3A	$a \wedge b \Leftrightarrow b \wedge a$	3B	$a \vee b \Leftrightarrow b \vee a$
4A	$(a \wedge b) \wedge c \Leftrightarrow a \wedge (b \wedge c)$	4B	$(a \vee b) \vee c \Leftrightarrow a \vee (b \vee c)$
5A	$a \wedge (b \vee c) \Leftrightarrow (a \wedge b) \vee (a \wedge c)$	5B	$a \vee (b \wedge c) \Leftrightarrow (a \vee b) \wedge (a \vee c)$
6A	$a' \wedge b' \Leftrightarrow (a \vee b)'$	6B	$a' \vee b' \Leftrightarrow (a \wedge b)'$

Some of these results are proved below. The rest are given as examples for the reader in the next exercise.

Example. Prove the results 1A, 2B, 3B, 4A, 5A above. 1A is almost self-evident. The truth of a and of a is the truth of a. 2B is equally simple. It is certainty that a statement a is true or false, but if a is truth, then a' is falsehood (and *vice versa*) hence $a \vee a' = 1$ (certainty). We could alternatively have proved these results by truth tables, which is the method we adopt for the remainder.

3B.

a	b	$a \vee b$	$b \vee a$
1	1	1	1
1	0	1	1
0	1	1	1
0	0	0	0

As all the results of columns 3 and 4 are the same, it follows that $a \vee b \Leftrightarrow b \vee a$.

4A.

a	b	c	$a \wedge b$	$b \wedge c$	$(a \wedge b) \wedge c$	$a \wedge (b \wedge c)$
1	1	1	1	1	1	1
1	0	1	0	0	0	0
1	1	0	1	0	0	0
1	0	0	0	0	0	0
0	1	1	0	1	0	0
0	0	1	0	0	0	0
0	1	0	0	0	0	0
0	0	0	0	0	0	0

In the above table, column 6 is found by combining columns 4 and 3; column 7 by combining columns 1 and 5. Hence

$$(a \wedge b) \wedge c = a \wedge (b \wedge c).$$

Extensions of result 4A can easily be derived, e.g.

$$(a \wedge b) \wedge c = (a \wedge c) \wedge b = (b \wedge a) \wedge c.$$

In fact, as in the corresponding result for sets, we can write $a \wedge b \wedge c$, and similarly $a \vee b \vee c$.

5A.

a	b	c	$b \vee c$	$a \wedge (b \vee c)$	$a \wedge b$	$a \wedge c$	$(a \wedge b) \vee (a \wedge c)$
1	1	1	1	1	1	1	1
1	0	1	1	1	0	1	1
1	1	0	1	1	1	0	1
1	0	0	0	0	0	0	0
0	1	1	1	0	0	0	0
0	0	1	1	0	0	0	0
0	1	0	1	0	0	0	0
0	0	0	0	0	0	0	0

We see that, in the above table, column 5 and column 8 are the same. Hence

$$a \wedge (b \vee c) \Leftrightarrow (a \wedge b) \vee (a \wedge c).$$

Exercise 32

The standard results to which reference is made are listed on p. 139. Prove that the following are valid (Nos. 1–6):

1. $a \vee a \Leftrightarrow a$ (Result 1*B*).

2. $a \wedge a' \Leftrightarrow 0$ (Result 2*A*).

3. $a \wedge b \Leftrightarrow b \wedge a$ (Result 3*A*).

4. $(a \vee b) \vee c \Leftrightarrow a \vee (b \vee c)$ (Result 4*B*).

5. $a \vee (b \wedge c) \Leftrightarrow (a \vee b) \wedge (a \vee c)$ (Result 5*B*).

6. $a' \vee b' \Leftrightarrow (a \wedge b)'$ (Result 6*B*).

7. Find whether or not

$$a \wedge (b \vee c) \Leftrightarrow (a \wedge b) \vee c.$$

5. The NOR Function

We now introduce the *nor* function $a_N b$, which is read as "neither a nor b". It is therefore $a' \wedge b'$, for any member of $a' \wedge b'$ is a member of the complement of a and of the complement of b and hence belongs neither to a nor to b. It follows that

$$a_N b \Leftrightarrow a' \wedge b'$$

$$\Leftrightarrow (a \vee b)', \text{ using an earlier result.}$$

Every logical function of a and b can be expressed entirely in terms of *nors* and this is of value when dealing with circuit theory. Two illustrations should suffice to explain the method.

Example. Show that $a \wedge b' \Leftrightarrow (a_N a)_N b$.

Proof: $a \wedge b' \Leftrightarrow a'_N b,$

for   both a and $b' \Leftrightarrow$ neither a' nor b.

Now how can we write a' as a *nor* function? If $a_N b$ means "neither a nor b", then $a_N a$ means "neither a nor a", i.e. it means a'. Hence

$$a \wedge b' \Leftrightarrow (a_N a)_N b.$$

Example. Show that

$$a' \vee b \Leftrightarrow [(a_N a)_N b]_N [(a_N a)_N b].$$

Proof: $a' \vee b \Leftrightarrow (a \wedge b')'$, from result 6B (p. 139),

on interchanging b and b'

$\Leftrightarrow [(a_N a)_N b]'$, from the example above

$\Leftrightarrow [(a_N a)_N b]_N [(a_N a)_N b],$

using the second step in the example above [viz., the idea that $k' \Leftrightarrow k_N k$, where in this case $k \Leftrightarrow (a_N a)_N b$].

The student can hardly have failed to notice that this short section on the *nor* function requires some careful thought, and the following suggestion should help him. One can convert directly from the *and* function to the *nor* function as in the first example above, but to convert from the *or* function to the *nor* function it is wise to convert firstly to the *and* function as in the second example above. Thus

$$and \rightarrow nor$$

$$or \rightarrow and \rightarrow nor,$$

are the steps of reasoning in each case. This is all very well, may be the student's next thought, but why introduce these peculiar *nor* function processes when the *and* and *or* functions seem to be so much more straightforward? The answer lies in the next chapter, on Logical Circuit Theory. We can either evolve circuits involving several kinds of basic unit or we can construct other circuits to do the same job, all the circuits in this case being virtually of the same basic pattern, i.e. *nor* circuits. We shall in fact construct both types, but the latter is of greater value for many practical purposes, although perhaps more tiresome to plan initially.

Exercise 33

1. Show that

(a) $a \vee b \Leftrightarrow (a' \wedge b')'$

(b) $a \vee b \Leftrightarrow (a_N b)'$

(c) $a \wedge b \Leftrightarrow a'_N b'$

2. Prove that

(a) $$a' \wedge b \Leftrightarrow a_N(b_N b)$$

(b) $$a \vee b' \Leftrightarrow [a_N(b_N b)]_N [a_N(b_N b)].$$

Express the following (Qns. 3–8) entirely as *nor* functions:

3. $a \wedge b$.

4. $a \vee b$.

5. $a \wedge (b \vee c)$.

6. $a' \vee b'$.

7. $a \vee (b \wedge c)$.

8. a.

CHAPTER 10

LOGICAL CIRCUIT THEORY

1. Abbreviated Notation for Symbolic Logic

It will be recalled that an abbreviated form of notation was introduced near the end of the Chapter 7, on sets. We now introduce an exactly analogous notation for logical operations.

We replace the *or* symbol of logic ($\vee$) by *addition*,

$$\text{i.e. } a \vee b \text{ becomes } a+b.$$

We replace the *and* symbol of logic ($\wedge$) by *multiplication*,

$$\text{i.e. } a \wedge b \text{ becomes } ab.$$

The notation for *nor* functions is explained on p. 141.
The symbol $\Leftrightarrow$ is replaced by the sign of equality ($=$).

Logic Notation	Abbreviated Notation
$a \vee a \Leftrightarrow a$	$a+a = a$
$a \vee b \Leftrightarrow b \vee a$	$a+b = b+a$
$a \vee (b \vee c) \Leftrightarrow (a \vee b) \vee c$	$a+(b+c) = (a+b)+c$
$a \wedge a \Leftrightarrow a$	$aa = a$
$a \wedge b \Leftrightarrow b \wedge a$	$ab = ba$
$a \wedge (b \wedge c) \Leftrightarrow (a \wedge b) \wedge c$	$a(bc) = (ab)c$
$a \wedge (b \vee c) \Leftrightarrow (a \wedge b) \vee (a \wedge c)$	$a(b+c) = ab+ac.$
$a \vee (b \wedge c) \Leftrightarrow (a \vee b) \wedge (a \vee c)$	$a+bc = (a+b)(a+c)$
$a \vee a' \Leftrightarrow 1$	$a+a' = 1$
$a \wedge a' \Leftrightarrow 0$	$aa' = 0$
$a' \wedge b' \Leftrightarrow (a \vee b)'$	$a'b' = (a+b)'$
$a' \vee b' \Leftrightarrow (a \wedge b)'$	$a'+b' = (ab)'$

With the exception that small letters are used in logic instead of capitals, which we use for sets, these results are identical with those obtained earlier in our abbreviated set notation. We shall use the results above for the remainder of this chapter. The notation for *nor* functions is unchanged.

2. AND and OR Circuits

Although Boolean algebra is of importance in the circuit theory of electronic computers it is not, unfortunately, the complete answer. It is, in the form of symbolic logic, adaptable to series and to parallel circuit layouts but not necessarily to other forms. We confine our studies in this book entirely to considerations where our earlier work can be applied.

Let us start with two simple switches A and B, joined firstly in series (Fig. 85) and secondly in parallel (Fig. 86). Now current either flows through A (when the switch is on) or it does not (when the switch is off). A similar situation applies with regard to switch B. Truth tables are constructed using assertions a and b, where

$$a = 1 \text{ (switch } A \text{ on)}, a = 0 \text{ (switch } A \text{ off)}$$
$$b = 1 \text{ (switch } B \text{ on)}, b = 0 \text{ (switch } B \text{ off)}.$$

In the series circuit, Fig. 85, current will flow *if and only if* A and B are both on, i.e. if $a = 1$, $b = 1$; this implies $ab = 1$ and corresponds to $AB = F$ (F indicating that current flows). The truth table gives the details:

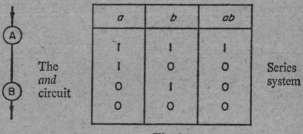

The *and* circuit

	a	b	ab
	1	1	1
	1	0	0
	0	1	0
	0	0	0

Series system

Fig. 85

This is clearly an *and* circuit.

In the parallel circuit, Fig. 86, current will flow if A is on *or* if B is on *or* if both A and B are on. This exactly implies $a + b = 1$, and corresponds to $A + B = F$ (defining F as above). The truth table is shown alongside the circuit.

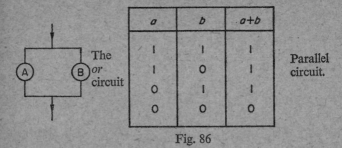

a	b	a+b
1	1	1
1	0	1
0	1	1
0	0	0

The *or* circuit

Parallel circuit.

Fig. 86

This is clearly an *or* circuit.

Complete electric circuits corresponding to the above switching systems are shown below (Figs. 87 and 88). Each has a battery, a lamp and two switches.

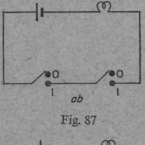

ab

Fig. 87

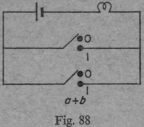

a+b

Fig. 88

A slightly more elaborate problem, and one found in many homes, is the circuit for switching on a lamp from either of two separate switches, e.g. a hall light operated by one switch upstairs and another downstairs. The circuit is shown in Fig. 89:

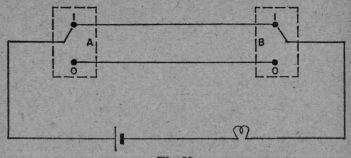

Fig. 89

Clearly a current will flow either if $a = 1$, $b = 1$ or if $a = 0$, $b = 0$. If we call $c = 1$ (current flowing) and $c = 0$ (no current), the truth table is

a	b	c
1	1	1
1	0	0
0	1	0
0	0	1

and c is formed of "ab and $a'b'$", i.e.

$$c = ab + a'b'.$$

Suppose now we cross the wires from A to B as shown in Fig. 90 below:

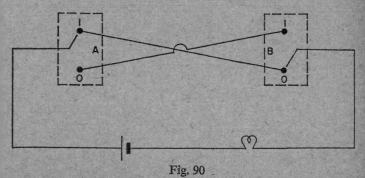

Fig. 90

The circuit is as effective as the previous one and does exactly the same job, but the truth table is different from that above, for its last column is the negation of the last column of the table above, i.e. the last column must give c' (the negation, or complement, of c):

a	b	c'
1	1	0
1	0	1
0	1	1
0	0	0

Here c' is formed of "ab' and $a'b$", i.e.

$$c' = ab' + a'b,$$

whence $\qquad\qquad c = (ab' + a'b)'.$

It therefore appears that

$$ab + a'b' = (ab' + a'b)'.$$

This can be shown to be true either by using truth tables for each of the expressions $ab + a'b'$, $(ab' + a'b)'$ or by using the theoretical results obtained earlier. It is left for the enthusiast to do this for himself.

Example. In the given circuit (Fig. 91), if $a = 1$, 0 corresponds to switch A being on, off respectively, etc., show that current will flow if

$$(a + b)c + d = 1.$$

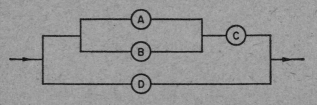

Fig. 91

Proof. Consider the upper part of the system. A, B in parallel are equivalent to $a+b$; bringing C into the calculation puts C in series with the A, B system. The upper part of the system therefore has a logical function $(a+b)c$. Finally, D is in parallel with the A, B, C circuit and so the whole network is equivalent to $(a+b)c+d$. Current flows when this is 1. Thus

$$(a+b)c+d = 1.$$

Exercise 34

Write down the logical functions for the following systems (Nos. 1–4):

1

Fig. 92
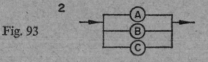

2

Fig. 93
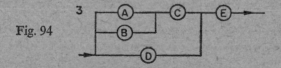

3

Fig. 94
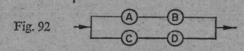

4

Fig. 95
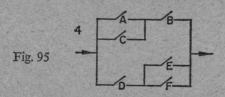

Use the notation given above.

5. In Nos. 1–4 above, given that $a = 1, b = 0, c = 1, d = 0, e = 1,$ $f = 0$, determine, in each exercise, whether a current will flow.
6. Repeat Qn. 5 above, given that $a = 0, b = 1, c = 0, d = 1, e = 0,$ $f = 1$.
7. Draw careful diagrams illustrating systems whose logical functions are (i) $(a+b)c$,

(ii) $ab+cde$,

(iii) $ab(c+de)$,

(iv) $(ab+c)(de+f)$,

(v) $ab'+a'b+c$,

(vi) $ab'+bc$.

3. The NOR Circuit

In section 2 above we considered two kinds of circuit, *and* and *or*. We need one more, which we call an *inverter*. This is one in which the output is the complement of the input, i.e. if the input is a the output is a'. We recall that

$$(a = 0) \Leftrightarrow (a' = 1); \quad (a = 1) \Leftrightarrow (a' = 0).$$

This implies that an input of current flowing leads to an output of no current and vice versa. The simple electromagnet, Fig. 96, serves our purpose. The whole device can be called an *inverting-gate*.

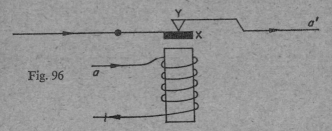

Fig. 96

If no input current a flows in the circuit around the electromagnet, contact is maintained along the output circuit a', i.e. $a = 0 \Leftrightarrow a' = 1$. If input current a flows around the electromagnet, the spring-mounted soft iron contact breaker X is drawn away from the point Y and contact in the output circuit is broken. Hence $a = 1 \Leftrightarrow a' = 0$. Thus both our conditions are satisfied.

We represent an inverter symbolically by

Fig. 97

On page 141 we showed that $a_N b \Leftrightarrow a' \wedge b'$, i.e. in our abbreviated notation

$$a_N b = a'b'.$$

In order, therefore, to provide a *nor* circuit $a_N b$ we need two inverters, the outputs being connected in series (Fig. 98).

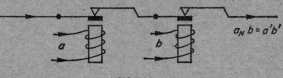

Fig. 98

Clearly current flows in the $a'b'$ (output) circuit if and only if $a = 0$ and $b = 0$. We represent such a circuit (called a *nor-gate*) symbolically by Fig. 99.

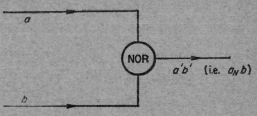

Fig. 99

The truth table for the nor-gate above is:

a	b	a'	b'	$a'b'$
1	1	0	0	0
1	0	0	1	0
0	1	1	0	0
0	0	1	1	1

We can obviously add more inverters in the same way, in a nor-gate, thereby getting a logical output $a_N b_N c \ldots = a'b'c' \ldots$. A truth table for three inputs is shown on page 154.

There is no real need to think of an inverter as a separate gate, for our *nor-gates* for one, two, three...inputs and one output are members of the same family (Fig. 100).

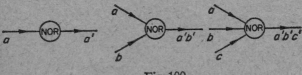

Fig. 100

It is now necessary to introduce two other gate symbols, the *and-gate* and the *or-gate*. The *and-gate* corresponds exactly to the series circuit ab (Fig. 85) and implies that output $ab = 1$ when both $a = 1$ and $b = 1$. Fig. 85 would not be of much value electrically for two inputs (or more) and Fig. 101 shows a modification of a nor-gate which will given an output when both (or all) currents are flowing round the electromagnets (i.e. it is an *and-gate*).

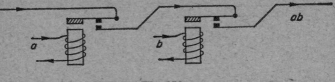

Fig. 101

The truth table has already been given on page 145. The symbol for an *and-gate* is shown below (Fig. 102).

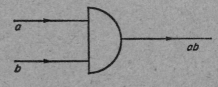

Fig. 102

The *or-gate* corresponds exactly to the parallel circuit $a+b$ (Fig. 86), but this figure also would not be of much practical interest. We can, however, once again modify our ideas on electromagnets to evolve a possible and workable scheme (Fig. 103). Only a very modest adjustment to the circuit in Fig. 101 is necessary to ensure an output from any one (or more) input $a, b \ldots$

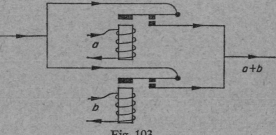

Fig. 103

The output $a+b = 1$ of an *or-gate* is implied when either $a = 1$ or $b = 1$ or both $a = 1$ and $b = 1$. The truth table has been given on page 146. The symbol for an or-gate is given below (Fig. 104).

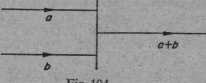

Fig. 104

As with the nor-gate, so we can add as many inputs as we wish to the and-gate and to the or-gate.

4. Logical Design: the Binary Half Adder

In the sections above, we have discussed systems with simple switches and electromagnets. For computing purposes these would be far too time-consuming[15]. On the other hand, the response of transistors to electric currents is very rapid indeed. Consequently, *and, or* and *nor* circuits containing transistors replace in electronic

---

[15] For tutorial purposes the electromagnetic schemes described can, however, be most instructive.

computers the simple devices we have so far considered. The actual equipment used is rather outside the scope of this book and is in fact irrelevant to the logical functions we are studying.

We are now familiar with three kinds of circuit gate. Suppose inputs $A, B, C \ldots$ whose logical functions are $a, b, c \ldots$ are fed into a circuit with a single output, then the possibilities are:

Circuit	Output
and	$abc\ldots$
or	$a+b+c\ldots$
nor	$a'b'c'\ldots$

Clearly the and-gate transmits an impulse if and only if all of $A, B, C \ldots$ feed impulses into it, i.e. if $a = b = c = \ldots = 1$.

The or-gate transmits an impulse if any one (or more) of $A, B, C \ldots$ feeds an impulse into it.

The nor-gate transmits an impulse if and only if none of $A, B, C \ldots$ feeds an impulse into it, i.e. if $a = b = c = \ldots = 0$. The following simple table (an extension of one on page 151) illustrates the case:

a	b	c	a'	b'	c'	$a_N b_N c$ $= a'b'c'$
1	1	1	0	0	0	0
1	1	0	0	0	1	0
1	0	1	0	1	0	0
0	1	1	1	0	0	0
1	0	0	0	1	1	0
0	1	0	1	0	1	0
0	0	1	1	1	0	0
0	0	0	1	1	1	1

Note that we have written down every possible combination of a, b, c. We have $a = 1$ or 0 and the same applies to b and c. There are thus two possibilities for each of the three functions a, b, c. Hence, from earlier work, there are $2^3 = 8$ combinations possible. There is obviously no output unless all the inputs are zero.

We are now in a position to consider the design of a *binary half-adder*, a machine designed to add two binary digits a and b (see Ch. 4, p. 63) either of which may be 0 or 1. The binary half-adder will give the unit and the carrying figure. Consider, in the binary scale,

$$1+1 = 10, \quad 1+0 = 1, \quad 0+1 = 1, \quad 0+0 = 0.$$

If we reset these as a truth table, we have

a	b	Addition	
		Carry	Unit
1	1	1	0
1	0	0	1
0	1	0	1
0	0	0	0

We deal with *carry* and *unit* separately. Firstly it is clear that the *carry* is given by *ab*. This requires a simple and-gate. The *unit* is much more subtle and is most easily seen by constructing a truth table, although the method of derivation may seem a little artificial.

a	b	ab'	a'b	ab'+a'b
1	1	0	0	0
1	0	1	0	1
0	1	0	1	1
0	0	0	0	0

The last column exactly agrees with the *unit* column of the previous table. Using *c* for *carry* and *u* for *unit*, we have

$$c = ab,$$
$$u = ab' + a'b.$$

Thus we get our binary half-adder (Fig. 105).

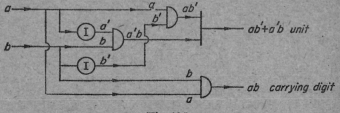

Fig. 105

This system uses six gates, viz. 2 inverters, 3 and-gates and 1 or-gate. The observant reader will have noticed that the upper part of the circuit giving the *unit* also answers the logical problem of the hall light (pages 146–8).

By suitable transformation, the system for the binary half-adder can be reduced from six gates to four (Fig. 106). The reasoning is as follows:

We have

$$u = ab' + a'b$$
$$= aa' + ab' + a'b + bb' \text{ (for } aa' = 0 = bb')$$
$$= (a+b)(a'+b')$$
$$= (a+b)(ab)'$$

This requires an or-gate (for $a+b$), an and-gate (for ab), an inverter (for $(ab)'$) and an and-gate to complete $(a+b)(ab)'$. No extra gate is needed for c, the carrying digit, for it is obtainable directly from the ab and-gate.

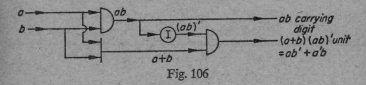

Fig. 106

5. The Whole Adder

The construction of a whole adder, which can add any two numbers (which are first converted to the binary scale) is effected by combining two half-adders.

The addition of two *numbers* requires that *three* digits shall be added together in each column, once the first step (in the units column) has been carried out. This follows at once from the fact that each number being added provides a digit and that there is also a carrying number (0 or 1) to be brought into each column, except the first. Hence a circuit designed to add two numbers needs three inputs. Such a device is termed a *whole adder* or *full adder*. Suppose that the inputs for a particular column are a, b and k (the carrying digit from the previous column).

We construct a truth table:

a	b	k	Total	Carry digit c	Unit digit u
0	0	0	0	0	0
0	0	1	1	0	1
0	1	0	1	0	1
1	0	0	1	0	1
0	1	1	10	1	0
1	0	1	10	1	0
1	1	0	10	1	0
1	1	1	11	1	1

We observe that the carry c is only needed when two or more of a, b, k are non-zero. This suggests

$$c = abk' + ab'k + a'bk + abk$$
$$= ab(k' + k) + (ab' + a'b)k$$
$$= ab + qk, \text{ where } q = ab' + a'b,$$

for $k + k' = 1$. The truth of our intuitive assumption is easily checked by the construction of a truth table for $ab + qk$.

We now note that the unit u is only non-zero when *one* or *three* of the inputs are non-zero. We surmise[16] therefore that

$$u = abk + ab'k' + a'bk' + a'b'k$$
$$= (ab + a'b')k + (ab' + a'b)k'$$
$$= q'k + qk',$$

for we have shown earlier (p. 148) that $ab + a'b' = (ab' + a'b)'$.

Finally, we observe that $ab' + a'b$, i.e. q, is the unit output of our half-adder with inputs a and b. Also $q'k + qk'$, i.e. u, is the unit output of the half-adder with inputs q and k.

We use the symbol

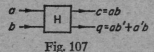

$$a \longrightarrow \boxed{H} \longrightarrow c = ab$$
$$b \longrightarrow \qquad \longrightarrow q = ab' + a'b$$

Fig. 107

[16] We can also check this by a truth table.

for an entire half-adder. Then the whole adder appears as Fig. 108.

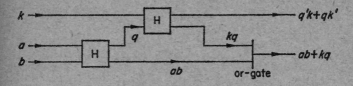

Fig. 108

Replacing our symbols for half-adders by the circuitry of Fig. 91, we have the complete picture for a whole adder (Fig. 109).

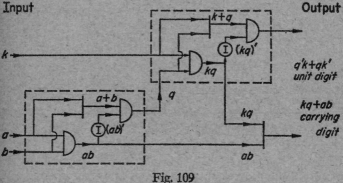

Fig. 109

Nine gates are needed in all. The half-adders are shown enclosed by dotted lines.

6. The NOR-gate Half Adder and Whole Adder

Although the foregoing sections 4 and 5 describe a satisfactory system of producing a binary half-adder and whole-adder, the circuits have the disadvantage that three different kinds of gate are needed. As the number of gates required in a computer may be very large, production and maintenance are much simplified if a single type of gate can be used. This is, in fact, the nor-gate system. Before, however, we can devise a logical circuit, we must convert our outputs for the *unit* and the *carrying* digit to a suitable nor-form. This is short but is not particularly easy.

From p. 155, we have the *carrying* digit for a half adder

$$c = ab \tag{1}$$

and this is in a form we can use, but the *unit*, obtained from p. 155, is much more tiresome. We start as follows:

$$u = (a+b)(ab)' \text{ (proved, p. 156)}$$
$$= (a'b')'(ab)',$$

for, from p. 144, $(a+b)' = a'b' \Leftrightarrow (a+b) = (a'b')'$.

Now, from p. 151, $p'q' = p_N q$; hence replacing p' by $(a'b')'$ and q' by $(ab)'$, we have

$$u = (a'b')_N(ab)$$
$$= (a_N b)_N(ab) \tag{2}$$

[Alternatively, we can say directly that "neither $a'b'$ nor ab" implies "both $(a'b')'$ and $(ab)'$", i.e. $(a'b')'(ab)'$.]

Unpromising as (1) and (2) may appear, they are exactly what we need.

We start by building up our half-adder.

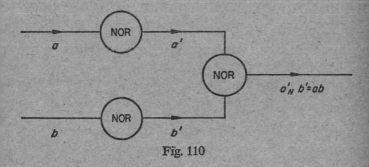

Fig. 110

This type of circuit (Fig. 110) not only gives the *carrying* digit but also part of the unit digit. The other part of the *unit* digit is $a_N b$, given by Fig. 111.

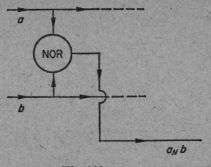

Fig. 111

Finally for the *unit*, we need one more nor-gate to give $(a_N b)_N(ab)$. Hence we have a complete half-adder in Fig. 112.

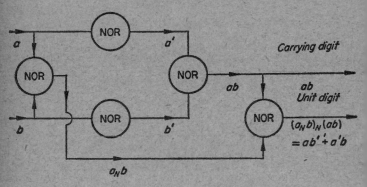

Fig. 112

Altogether five nor-gates are needed for the half-adder.

We now extend the construction of the nor-circuits to a binary whole adder. The principle is exactly that employed in section 5 above for the connection of two binary half adders. We start by redrawing Fig. 108 and use our half-adder symbol (Fig. 107), thus obtaining Fig. 113.

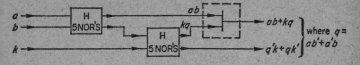

Fig. 113

To complete our problem we must convert our or-gate (enclosed in dotted lines (Fig. 113) to nor-gates. This is effected in Fig. 114,

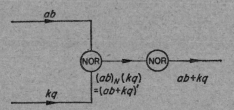

Fig. 114

for the first nor-gate converts inputs ab and kq to

$$(ab)_N(kq) = (ab)'(kq)' = (ab+kq)',$$

and the second nor-gate converts this to $ab+kq$. We thus have the whole adder (Fig. 115):

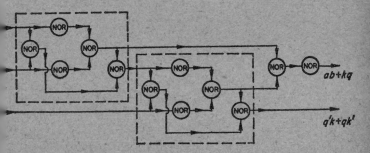

Fig. 115

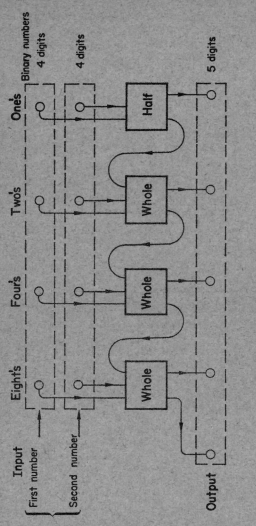

Fig 116

Thus, for a whole adder, 12 nor-gates are needed, compared with the 9 assorted gates of Fig. 109. Nevertheless, Fig. 115 presents a system of considerable interest, as has already been mentioned.

7. Binary Addition Computer

We complete our studies in circuit theory by drawing a system for adding two binary numbers with not more than 4 digits in each. It is self-evident that the system may be indefinitely extended (Fig. 116).

The one's column only needs a half-adder, as there is no carrying figure to be brought in. All the other columns need whole-adders. The machine would add two such numbers as $1110 + 1101$ and give the answer as 11011.

Exercise 35

1. Prove, by constructing truth tables, the formulae:

      (1) $u = abk + ab'k' + a'bk' + a'b'k,$

      (2) $c = abk' + ab'k + a'bk + abk,$

for the *unit* digit, *carrying* digit, respectively, when three binary digits a, b, k are added together (p. 157).

2. Show that, for the computer in Fig. 116, the largest input numbers expressed in the denary scale are each 15, and that the largest output is 30. Why cannot 31 be obtained on this particular instrument?

3. (*a*) Which figure shows exactly the construction of an and-gate using nor-gates only? (*b*) Draw a figure for an or-gate, using nor-gates each having two inputs and one output.

4. Draw from memory the figures for
    (*a*) a binary half-adder using 6 gates,
    (*b*) a binary half-adder using nor-gates only,
    (*c*) a whole adder using nor-gates only.

5. Prove, by constructing truth tables or otherwise, that
    (*a*) $a + b = (a_N b)_N (a_N b)$
    (*b*) $(a + b)' = a_N b$
    (*c*) $ab' + a'b = (a + b)(ab)'.$

CHAPTER 11

MATRICES

1. Matrices

Suppose we consider two simultaneous linear equations in two unknowns x and y,

$$3x+y = 5$$
$$4x-y = 2$$

The coefficients of x and y could be put down in the form of a rectangular array without altering their relative positions in the equations, thus:

$$\begin{pmatrix} 3 & 1 \\ 4 & -1 \end{pmatrix}$$

Such an array is called a *matrix*. We can, in fact, write the pair of equations as a single matrix equation

$$\begin{pmatrix} 3 & 1 \\ 4 & -1 \end{pmatrix}\begin{pmatrix} x \\ y \end{pmatrix} = \begin{pmatrix} 5 \\ 2 \end{pmatrix}$$

This suggests the *raison d'être* for matrix algebra. In aerodynamics, for example, large numbers of simultaneous linear equations involving a large number of unknowns can be replaced by a single matrix equation. This would, however, be of little value unless we could find rules and methods for operating with matrices. We accordingly begin by examining properties of these arrays.

Definition. A matrix is a rectangular array of numbers. An m by n matrix (written $m \times n$ matrix) has m rows and n columns, e.g.

$$\begin{pmatrix} 2 & 4 & 7 \\ 1 & 6 & 2 \end{pmatrix} \qquad \text{is a } 2 \times 3 \text{ matrix} \qquad \text{(i)}$$

$$\begin{pmatrix} 5 & 2 & 7 \\ -1 & 0 & 3 \\ 3 & 4 & 1 \end{pmatrix} \qquad \text{is a square } 3 \times 3 \text{ matrix} \qquad \text{(ii)}$$

$$\begin{pmatrix} 2 & 5 \\ 0 & 1 \\ 1 & 0 \end{pmatrix} \qquad \text{is a } 3 \times 2 \text{ matrix} \qquad \text{(iii)}$$

$$(1 \quad 2 \quad 9) \qquad \text{is a row } 1 \times 3 \text{ matrix} \qquad \text{(iv)}$$

$$\begin{pmatrix} 4 \\ 2 \\ 3 \end{pmatrix} \qquad \text{is a column } 3 \times 1 \text{ matrix} \qquad \text{(v)}$$

We are already familiar with the idea of laying out rectangular patterns in mathematics, e.g. $2 \times 3 = 6$ can be represented as

$$\begin{matrix} \bullet & \bullet & \bullet \\ \bullet & \bullet & \bullet \end{matrix}$$

The analogy in matrix algebra is close, for a 2×3 matrix has 6 elements (c.f. matrix (i) above). Likewise a 3×2 matrix has 6 elements (c.f. matrix (iii) above) but this time the elements are arranged differently.

In general, a matrix can be written

$$\begin{pmatrix} a_{11} & a_{12} & a_{13} & \cdots & a_{1n} \\ a_{21} & a_{22} & a_{23} & \cdots & a_{2n} \\ \cdot & \cdot & \cdot & \cdot & \cdot \\ a_{m1} & a_{m2} & a_{m3} & \cdots & a_{mn} \end{pmatrix}$$

and this is sometimes written symbolically (a_{mn}), but the investigation of the general case requires some skill in mathematics and is rather beyond the scope of this book. We shall restrict our investigations to matrices no greater than 3×3, except where reference to the general case presents no difficulty. Clarendon type letters are usually used to represent matrices, e.g. three matrices could be represented as $\mathbf{A}, \mathbf{B}, \mathbf{C}$.

2. Simple Matrix Properties

The position of an element in a matrix is of fundamental importance. If different elements are interchanged, the matrix itself is changed. For example,

$$\begin{pmatrix} 2 & 1 \\ 5 & 7 \end{pmatrix}, \quad \begin{pmatrix} 1 & 2 \\ 5 & 7 \end{pmatrix}, \quad \begin{pmatrix} 2 & 5 \\ 1 & 7 \end{pmatrix}$$

are all different.

Two matrices are identical if and only if each element of one is equal to the corresponding element of the other. Using the symbolic notation of Section 1 above, if

$$\mathbf{A} = (a_{mn}) \quad \text{and} \quad \mathbf{B} = (b_{mn})$$

then

$$\mathbf{A} = \mathbf{B} \Leftrightarrow (a_{mn}) = (b_{mn})$$

$$\Leftrightarrow a_{rs} = b_{rs},$$

for all $1 \leqslant r \leqslant m$, $1 \leqslant s \leqslant n$, where r, s, m, n are integers.

For instance,

$$\begin{pmatrix} a_{11} & a_{12} & a_{13} \\ a_{21} & a_{22} & a_{23} \end{pmatrix} = \begin{pmatrix} b_{11} & b_{12} & b_{13} \\ b_{21} & b_{22} & b_{23} \end{pmatrix}$$

if and only if

$$a_{11} = b_{11}, a_{12} = b_{12}, a_{13} = b_{13}, a_{21} = b_{21}, a_{22} = b_{22}, a_{23} = b_{23}.$$

Further, for example,

$$\begin{pmatrix} x & 4 \\ 3 & y \end{pmatrix} = \begin{pmatrix} 1 & 4 \\ 3 & 2 \end{pmatrix}$$

$$\Leftrightarrow x = 1 \text{ and } y = 2.$$

Two matrices cannot be equal unless they have the same number of rows (m, say) and the same number of columns (n, say). Thus

$$\begin{pmatrix} 2 & 0 \\ -3 & 0 \end{pmatrix} \neq \begin{pmatrix} 2 \\ -3 \end{pmatrix},$$

the symbol $\neq$ meaning "does not equal".

The *addition* (or *subtraction*) of matrices is simple but may only be carried out when the matrices are of the same order ($m \times n$, say), i.e. when they have the same number of rows and columns. Corresponding elements are added (or, in the case of subtraction, the elements are subtracted).

Example. $(4 \quad -3) + (-2 \quad 6) + (1 \quad 0) = (4-2+1 \quad -3+6+0)$
$$= (3 \quad 3).$$

Example.
$$\begin{pmatrix} 4 & 2 \\ -1 & 0 \\ 0 & 5 \end{pmatrix} + \begin{pmatrix} -2 & 1 \\ 1 & 5 \\ 3 & -5 \end{pmatrix} = \begin{pmatrix} 4-2 & 2+1 \\ -1+1 & 0+5 \\ 0+3 & 5-5 \end{pmatrix}$$

$$= \begin{pmatrix} 2 & 3 \\ 0 & 5 \\ 3 & 0 \end{pmatrix}.$$

Intermediate steps may be omitted after a little practice.

We could *not* add

$$\begin{pmatrix} 6 & 1 \\ 0 & 4 \end{pmatrix} \quad \text{and} \quad \begin{pmatrix} 3 \\ 8 \end{pmatrix}$$

as, although the number of rows is the same (2), the number of columns is not (2 in the first matrix, 1 in the second).

If matrices *may* be added, we refer to them in the theorem below as compatible.

Theorem. The addition of compatible matrices is commutative, i.e. $A+B = B+A$.

We prove this for a general 2×2 matrix.

Suppose

$$A = \begin{pmatrix} a_{11} & a_{12} \\ a_{21} & a_{22} \end{pmatrix} \quad \text{and} \quad B = \begin{pmatrix} b_{11} & b_{12} \\ b_{21} & b_{22} \end{pmatrix}$$

then

$$A+B = \begin{pmatrix} a_{11} & a_{12} \\ a_{21} & a_{22} \end{pmatrix} + \begin{pmatrix} b_{11} & b_{12} \\ b_{21} & b_{22} \end{pmatrix}$$

$$= \begin{pmatrix} a_{11}+b_{11} & a_{12}+b_{12} \\ a_{21}+b_{21} & a_{22}+b_{22} \end{pmatrix}.$$

Now

$$B+A = \begin{pmatrix} b_{11} & b_{12} \\ b_{21} & b_{22} \end{pmatrix} + \begin{pmatrix} a_{11} & a_{12} \\ a_{21} & a_{22} \end{pmatrix}$$

$$= \begin{pmatrix} b_{11}+a_{11} & b_{12}+a_{12} \\ b_{21}+a_{21} & b_{22}+a_{22} \end{pmatrix}$$

$$= \begin{pmatrix} a_{11}+b_{11} & a_{12}+b_{12} \\ a_{21}+b_{21} & a_{22}+b_{22} \end{pmatrix} \qquad \text{(for each element is unchanged)}$$

$$= A+B.$$

It is obvious that we can proceed in exactly the same way in handling any two $m \times n$ matrices, and that the process can be extended to the addition of three or more matrices, e.g.

$$\mathbf{A}+\mathbf{B}+\mathbf{C} = \mathbf{B}+\mathbf{A}+\mathbf{C} = \mathbf{B}+\mathbf{C}+\mathbf{A}, \text{ etc.}$$

Example. $\begin{pmatrix} x^2+2x-5 \\ y^2+1 \end{pmatrix} + \begin{pmatrix} 5x+5 \\ 2-y \end{pmatrix} = \begin{pmatrix} x^2+7x \\ y^2-y+3 \end{pmatrix}.$

Observe that these are column matrices and that the fact that the elements are algebraic expressions does not affect the rules of matrices.

A matrix having every element zero is called a zero matrix. If, to a given matrix A, a zero matrix is added, the value of the former is unchanged, e.g.

$$\begin{pmatrix} 3 & 5 & 0 \\ 2 & 0 & 1 \end{pmatrix} + \begin{pmatrix} 0 & 0 & 0 \\ 0 & 0 & 0 \end{pmatrix} = \begin{pmatrix} 3 & 5 & 0 \\ 2 & 0 & 1 \end{pmatrix}$$

but, as in the work above, our zero matrix must, to be compatible, that is (in the case of addition or subtraction), it must have the same number of rows and columns as the matrix to which we add it. In general,

$$\mathbf{A}+\phi = \mathbf{A},$$

where A is an $m \times n$ matrix and ϕ is the corresponding $m \times n$ zero matrix.

Exercise 36

1. Write down examples of
   - (a) a square 2×2 matrix,
   - (b) a 1×3 (row) matrix,
   - (c) a 2×1 (column) matrix,
   - (d) a 3×3 square matrix,
   - (e) a 2×3 matrix.
   - (f) a 3×2 matrix.

[*Hint.* An example of a 1×3 (row) matrix would be $(-2 \quad 0 \quad 5)$.]

2. If the matrix A were expressed by the shortened notation $\mathbf{A} = (a_{mn})$ how many rows and columns would there be if $m = 5$, $n = 3$?

3. State whether the following can be added. In all cases where permissible carry out possible simplification:

(a) $\begin{pmatrix} 2 & 1 \\ -3 & 0 \end{pmatrix} + \begin{pmatrix} 0 & 4 \\ 3 & 2 \end{pmatrix}$, (b) $\begin{pmatrix} 3 \\ 1 \\ 0 \end{pmatrix} + \begin{pmatrix} 0 & 1 \\ 1 & 0 \\ 2 & 4 \end{pmatrix}$,

(c) $(3 \quad 1 \quad -4) - (-2 \quad -1 \quad 3) + (-5 \quad 0 \quad 2),$

(d) $(4 \quad 2) + (3 \quad -7) - (0),$ (e) $\begin{pmatrix} -4 \\ 2 \end{pmatrix} - \begin{pmatrix} -3 \\ 3 \end{pmatrix}$

(f) $\begin{pmatrix} 2 & 3 \\ 6 & -2 \\ -5 & 0 \end{pmatrix} + \begin{pmatrix} 4 & 1 \\ 4 & 2 \\ 5 & 3 \end{pmatrix},$ (g) $\begin{pmatrix} 6 & 1 & 7 \\ -3 & 2 & 5 \\ 4 & -1 & 8 \end{pmatrix} - \begin{pmatrix} 4 & -1 & -2 \\ 5 & 0 & -3 \\ -3 & -1 & 4 \end{pmatrix}$

4. From the following matrix equations, determine the values of x and y:

(a) $\begin{pmatrix} 2x & 5 \\ 3 & -1 \end{pmatrix} = \begin{pmatrix} -4 & 5 \\ 3 & -1 \end{pmatrix},$ (b) $\begin{pmatrix} 2 & 3x \\ y^2 & 1 \end{pmatrix} = \begin{pmatrix} 2 & 2 \\ -4 & 1 \end{pmatrix},$

(c) $\begin{pmatrix} 3x+2 & 7 \\ -2 & 0 \end{pmatrix} = \begin{pmatrix} y-1 & 7 \\ -2 & y \end{pmatrix},$ (d) $\begin{pmatrix} x+2y & 14 \\ -3 & y-2 \end{pmatrix} = \begin{pmatrix} 4 & 14 \\ -3 & 7+3x \end{pmatrix}.$

(e) $\begin{pmatrix} 1 & x^2+12 \\ x+4 & 3 \end{pmatrix} = \begin{pmatrix} 1 & 7x \\ x+4 & 3 \end{pmatrix}.$

5. Can we find x and y in the following matrix equation?

$$\begin{pmatrix} 4x^2 & 5 \\ 3 & y+2 \end{pmatrix} = \begin{pmatrix} 9 & 5 \\ 3 & y+2 \end{pmatrix}.$$

3. Multiplication of Matrices by Real Numbers

Consider

$$\begin{pmatrix} 2 & 3 \\ -1 & 2 \end{pmatrix} + \begin{pmatrix} 2 & 3 \\ -1 & 2 \end{pmatrix} = \begin{pmatrix} 4 & 6 \\ -2 & 4 \end{pmatrix}$$

using the rules of Sections 1 and 2. Now the matrices on the left hand sides of the equation are the same, so we can write

$$2\begin{pmatrix} 2 & 3 \\ -1 & 2 \end{pmatrix} = \begin{pmatrix} 4 & 6 \\ -2 & 4 \end{pmatrix}.$$

The result is perfectly general, although rather more tedious to prove, i.e. multiplying a matrix by a number p, say, multiplies every element of the matrix by p. Thus, in our general form already used, if $A = (a_{mn})$, then $pA = p(a_{mn}) = (pa_{mn})$.

Written in full this is

$$p\begin{pmatrix} a_{11} & a_{12} & \cdots & a_{1n} \\ a_{21} & a_{22} & \cdots & a_{2n} \\ \cdot & \cdot & \cdot & \cdot \\ a_{m1} & a_{m2} & \cdots & a_{mn} \end{pmatrix} = \begin{pmatrix} pa_{11} & pa_{12} & \cdots & pa_{1n} \\ pa_{21} & pa_{22} & \cdots & pa_{2n} \\ \cdot & \cdot & \cdot & \cdot \\ pa_{m1} & pa_{m2} & \cdots & pa_{mn} \end{pmatrix}$$

We can think of subtraction as multiplying the relevant matrix by -1 and then the adding it, e.g.

$$\begin{pmatrix} 3 & 5 \\ -4 & 7 \end{pmatrix} - \begin{pmatrix} 2 & 5 \\ -4 & 7 \end{pmatrix} = \begin{pmatrix} 3 & 5 \\ -4 & 7 \end{pmatrix} + [-1]\begin{pmatrix} 2 & 5 \\ -4 & 7 \end{pmatrix}$$

$$= \begin{pmatrix} 3 & 5 \\ -4 & 7 \end{pmatrix} + \begin{pmatrix} -2 & -5 \\ 4 & -7 \end{pmatrix}$$

$$= \begin{pmatrix} 1 & 0 \\ 0 & 0 \end{pmatrix},$$

and this agrees with the rule for subtraction given in Section 2 above. Note that we put the multiplier -1 in a square bracket to avoid comparison with (-1) which, in a chapter on matrices, should mean a 1×1 matrix.

Example. If $A = \begin{pmatrix} 2 & 1 & 4 \\ -1 & 3 & 0 \end{pmatrix}$ and $B = \begin{pmatrix} 1 & 2 & 5 \\ 3 & 0 & -1 \end{pmatrix}$,

find $2A - 3B$.

We have $2A - 3B = \begin{pmatrix} 4 & 2 & 8 \\ -2 & 6 & 0 \end{pmatrix} - \begin{pmatrix} 3 & 6 & 15 \\ 9 & 0 & -3 \end{pmatrix}$

$$= \begin{pmatrix} 1 & -4 & -7 \\ -11 & 6 & 3 \end{pmatrix}.$$

Definition. A symmetrical matrix is a square matrix (it cannot be any other kind) which is unaltered when rows and columns are interchanged; i.e.

$$\begin{pmatrix} 1 & 4 & -2 \\ 4 & 3 & 1 \\ -2 & 1 & 0 \end{pmatrix} \text{ is a symmetrical matrix.}$$

In general the square matrix (a_{nn}) is symmetrical if $a_{rs} = a_{sr}$, when $1 \leqslant r \leqslant n, 1 \leqslant s \leqslant n$.

Exercise 37

Multiply out (Nos. 1–4) and simplify where possible:

1. $5 \begin{pmatrix} 1 & 2 \\ 3 & -4 \\ 0 & 5 \end{pmatrix}$.

2. $-4 \begin{pmatrix} -1 & 0 & 4 \\ 2 & 1 & -3 \\ -2 & 5 & 0 \end{pmatrix}$.

3. $2 \begin{pmatrix} 4 & 1 & 7 \\ -2 & -3 & 0 \end{pmatrix} + \begin{pmatrix} -8 & 0 & -2 \\ 4 & 6 & 0 \end{pmatrix}$.

4. $3 \begin{pmatrix} 6 & 0 & 2 \\ 4 & 0 & 5 \\ 3 & 8 & 1 \end{pmatrix} - 2 \begin{pmatrix} 9 & 0 & 3 \\ 6 & 0 & 5 \\ 4 & 12 & 0 \end{pmatrix}$.

5. If

$$\mathbf{A} = \begin{pmatrix} x & 3 & -7 \\ 4 & 0 & 6y \end{pmatrix} \text{ and } \mathbf{B} = \begin{pmatrix} 1-y & 9 & -21 \\ 12 & 0 & 36 \end{pmatrix},$$

and if $3\mathbf{A} = \mathbf{B}$, find x and y.

6. Express the following as multiples of simpler matrices:

$$(a) \begin{pmatrix} 17 & 51 \\ -34 & 0 \end{pmatrix}, \quad (b) \begin{pmatrix} 91 & 63 & -56 \\ 49 & 0 & -98 \end{pmatrix}.$$

4. Multiplication of Matrices

The *multiplication* of matrices has an important place in modern mathematics. It is, however, more exacting than the process of *addition* and the student will need to observe some care in studying this section. Once the procedure is mastered, it is not difficult to apply.

Consider this domestic problem.

Mrs. Newton who, *mirabile dictu*, has a mathematical turn of mind, buys some groceries, namely, 3 lb. butter, 1 lb. tea, 2 lb. cheese. She lays out her purchases as a row matrix:

$$(3 \quad 1 \quad 2).$$

The costs of her groceries are based on the following prices per lb., say,

$$
\begin{array}{ll}
\text{butter} & \text{15p} \\
\text{tea} & \text{30p} \\
\text{cheese} & \text{20p.}
\end{array}
$$

These she puts down as a column matrix

$$
\begin{pmatrix} 15 \\ 30 \\ 20 \end{pmatrix}
$$

and the whole transaction is the product of the two matrices Mrs. Newton has constructed, namely,

$$
\begin{pmatrix} 3 & 1 & 2 \end{pmatrix} \begin{pmatrix} 15 \\ 30 \\ 20 \end{pmatrix}.
$$

We know already, from elementary arithmetic, that the total cost in pence is the sum of the products of each element in the first matrix row with the corresponding element in the second matrix column, i.e.

$$
(3 \times 15 + 1 \times 30 + 2 \times 20) = (115)
$$

which gives a bill of £1·15. Doubtless Mr. Ramsbottom, the practically minded grocer whose gifts of mathematical reasoning are limited, will have done his calculation in this way, without even realising that he was resorting to matrix algebra!

Suppose, like Mrs. Newton, we speculate as to whether we can extend the multiplication of elements of rows of the first matrix by elements of columns of the second matrix in a pair of matrices being multiplied together. The following could follow logically: if A and **B** are compatible matrices (and compatibility in multiplication has yet to be determined), then

$$
\mathbf{A} \times \mathbf{B} = \begin{pmatrix} \text{first row} \times \text{first column, first row} \times \text{second column,} \dots \\ \text{second row} \times \text{first column, second row} \times \text{second column,} \\ \cdots\cdots\cdots\cdots\cdots\cdots\cdots\cdots\cdots \end{pmatrix}
$$

Example. $\begin{pmatrix} 4 & -2 \\ -1 & 0 \end{pmatrix}\begin{pmatrix} 3 & 1 \\ 2 & 5 \end{pmatrix} = \begin{pmatrix} 4\times3-2\times2 & 4\times1-2\times5 \\ -1\times3+0\times2 & -1\times1+0\times5 \end{pmatrix}$

$$= \begin{pmatrix} 8 & -6 \\ -3 & -1 \end{pmatrix}.$$

More generally, for a 2×2 (square) matrix

$$\begin{pmatrix} a_{11} & a_{12} \\ a_{21} & a_{22} \end{pmatrix}\begin{pmatrix} b_{11} & b_{12} \\ b_{21} & b_{22} \end{pmatrix} = \begin{pmatrix} a_{11}b_{11}+a_{12}b_{21} & a_{11}b_{12}+a_{12}b_{22} \\ a_{21}b_{11}+a_{22}b_{21} & a_{21}b_{12}+a_{22}b_{22} \end{pmatrix}.$$

In words, the procedure is to multiply each element in the *first row* of the *first* matrix by the corresponding element in the *first column* of the second matrix and to *add* the results, thereby obtaining the *first element* of the *first row* in the final matrix. The routine is applied again using the *first row* of the *first* matrix and the *second column* of the *second* matrix, thereby getting the *second* element of the *first* row in the final matrix. Having completed the first row of the final matrix, we begin again using the *second* row of the first matrix and the *first* column of the second matrix to obtain the *first* element of the *second* row of the final matrix, and so on.

It sounds terrible, but it is not really difficult. One or two more examples should clarify the method:

Example. $\begin{pmatrix} 2 & 1 \\ 5 & 4 \end{pmatrix}\begin{pmatrix} -1 \\ 3 \end{pmatrix} = \begin{pmatrix} -2+3 \\ -5+12 \end{pmatrix} = \begin{pmatrix} 1 \\ 7 \end{pmatrix}$

Example. $\begin{pmatrix} 4 \\ -2 \end{pmatrix}(3 \quad 0 \quad 1) = \begin{pmatrix} 12 & 0 & 4 \\ -6 & 0 & -2 \end{pmatrix}$

$$= 2\begin{pmatrix} 6 & 0 & 2 \\ -3 & 0 & -1 \end{pmatrix}.$$

The general case for a 3×3 matrix is given in full below:

$$\begin{pmatrix} a_{11} & a_{12} & a_{13} \\ a_{21} & a_{22} & a_{23} \\ a_{31} & a_{32} & a_{33} \end{pmatrix}\begin{pmatrix} b_{11} & b_{12} & b_{13} \\ b_{21} & b_{22} & b_{23} \\ b_{31} & b_{32} & b_{33} \end{pmatrix}$$

$$\begin{pmatrix} a_{11}b_{11}+a_{12}b_{21} & a_{11}b_{12}+a_{12}b_{22} & a_{11}b_{13}+a_{12}b_{23} \\ \quad +a_{13}b_{31} & +a_{13}b_{32} & +a_{13}b_{33} \\[1em] a_{21}b_{11}+a_{22}b_{21} & a_{21}b_{12}+a_{22}b_{22} & a_{21}b_{13}+a_{22}b_{23} \\ \quad +a_{23}b_{31} & +a_{23}b_{32} & +a_{23}b_{33} \\[1em] a_{31}b_{11}+a_{32}b_{21} & a_{31}b_{12}+a_{32}b_{22} & a_{31}b_{13}+a_{32}b_{23} \\ \quad +a_{33}b_{31} & +a_{33}b_{32} & +a_{33}b_{33} \end{pmatrix}$$

We have so far not considered whether there are any snags in all this. All was well whilst we were multiplying the above matrices, but there was an essential rule which we carefully observed (but did not at that stage mention):

Rule. Two matrices may be multiplied together if and only if the number of columns of the first matrix is equal to the number of rows of the second matrix.

In other words, if A and B are two matrices being multiplied together in the order $A \times B$, then the result has no meaning unless A is an $m \times p$ matrix (say) and B is a $p \times n$ matrix (say), where p is the same in both cases. When this condition is satisfied the resultant matrix C is an $m \times n$ matrix. Symbolically, we can write

$$(a_{mp}) \times (b_{pn}) = (c_{mn}),$$

where, of course, every "c" has to be determined.

N.B. It does not follow that $B \times A$ exists and, if it does, it is most unlikely to be the same as $A \times B$. If $A = (a_{mn})$, $B = (b_{nm})$ then $A \times B$ *and* $B \times A$ exist, $A \times B$ being an $m \times m$ square matrix and $B \times A$ being an $n \times n$ square matrix.

Example. Given that

$$A = \begin{pmatrix} 1 & 0 & -1 \\ 0 & 5 & 3 \end{pmatrix}, \qquad B = \begin{pmatrix} 1 & 3 \\ -2 & 0 \\ 4 & 6 \end{pmatrix},$$

find $A \times B$ and $B \times A$. What can be deduced from these results?

Firstly we observe that A is a 2×3 matrix and that B is a 3×2 matrix, so

$$A \times B = (a_{23}) \times (b_{32}) = (c_{22}) \Rightarrow 2 \times 2 \text{ matrix}$$

and

$$B \times A = (b_{32}) \times (a_{23}) = (d_{33}) \Rightarrow 3 \times 3 \text{ matrix},$$

where the c's and d's have to be determined.

It follows at once that $\mathbf{A} \times \mathbf{B} \neq \mathbf{B} \times \mathbf{A}$, i.e. in general the multiplication of matrices is not commutative. This is the deduction for which we were asked.

Finally

$$\mathbf{A} \times \mathbf{B} = \begin{pmatrix} 1 & 0 & -1 \\ 0 & 5 & 3 \end{pmatrix} \begin{pmatrix} 1 & 3 \\ -2 & 0 \\ 4 & 6 \end{pmatrix} = \begin{pmatrix} -3 & -3 \\ 2 & 18 \end{pmatrix}$$

and

$$\mathbf{B} \times \mathbf{A} = \begin{pmatrix} 1 & 3 \\ -2 & 0 \\ 4 & 6 \end{pmatrix} \begin{pmatrix} 1 & 0 & -1 \\ 0 & 5 & 3 \end{pmatrix} = \begin{pmatrix} 1 & 15 & 8 \\ -2 & 0 & 2 \\ 4 & 30 & 14 \end{pmatrix}.$$

Note. Slightly different notations exist for multiplication of matrices, i.e. $\mathbf{A} \times \mathbf{B}$, $\mathbf{A}.\mathbf{B}$ and $\mathbf{AB}$, but in all cases the *order* of the matrices is preserved. [Thus, $\mathbf{A} \times \mathbf{B} = \mathbf{AB} \neq \mathbf{BA}$.]

Exercise 38

1. State which of the following matrix multiplications are permissible. Wherever possible find the resultant matrix:

(a) $(5)(-3)$,

(b) $(2, -1)\begin{pmatrix} 0 \\ 4 \end{pmatrix}$,

(c) $(3, 2)\begin{pmatrix} 2 & 5 \\ 0 & -1 \end{pmatrix}$,

(d) $\begin{pmatrix} 2 \\ 1 \end{pmatrix}(4 \quad 3)$,

(e) $\begin{pmatrix} 3 \\ -2 \end{pmatrix}\begin{pmatrix} 2 & 0 \\ 4 & -3 \end{pmatrix}$,

(f) $(3 \quad 0 \quad -5)\begin{pmatrix} 2 \\ 1 \\ -1 \end{pmatrix}$,

(g) $\begin{pmatrix} 2 \\ 0 \\ -3 \end{pmatrix}(5 \quad 4)$,

(h) $\begin{pmatrix} 4 \\ 1 \\ 2 \end{pmatrix}(3)$,

(i) $\begin{pmatrix} 2 & 0 \\ 5 & -3 \\ 0 & 1 \end{pmatrix}\begin{pmatrix} 1 & 5 \\ -3 & 2 \end{pmatrix}$.

2. If $\mathbf{A} = (2 \quad 1 \quad 0)$, $\mathbf{B} = \begin{pmatrix} 2 \\ 4 \\ 3 \end{pmatrix}$, find $\mathbf{AB}$ and $\mathbf{BA}$.

3. Given that $\mathbf{P} = \begin{pmatrix} 2 & 1 \\ 3 & 5 \end{pmatrix}$, $\mathbf{Q} = \begin{pmatrix} 5 \\ -4 \end{pmatrix}$, do $\mathbf{PQ}$ and $\mathbf{QP}$ have any meaning? If possible, express the result (or results) as a single matrix (or as matrices).

4. Simplify $\begin{pmatrix} 3 & 6 & -2 \\ 0 & 1 & -1 \\ -4 & 2 & 5 \end{pmatrix} \begin{pmatrix} 0 & 1 & 3 \\ 2 & 0 & 5 \\ 3 & 3 & -1 \end{pmatrix}$.

5. If $\begin{pmatrix} x & a & b \\ c & d & y \end{pmatrix} = \begin{pmatrix} 4 \\ 5 \end{pmatrix} (3 \quad 0 \quad -2)$, find x and y.

[*Hint.* We do not need to consider the values of a, b, c, d here, although if desired we could find them at once.]

6. Find $3 \begin{pmatrix} 2 & 4 & 0 \\ -1 & 3 & 2 \end{pmatrix} + 2 \begin{pmatrix} -3 & 0 & 1 \\ 2 & 5 & -3 \end{pmatrix}$.

7. Determine a_{32} and a_{21}, when

$$A = \begin{pmatrix} 3 & 1 & -2 \\ 0 & 4 & 3 \\ 2 & 0 & 1 \end{pmatrix} \quad \text{and} \quad B = \begin{pmatrix} 0 & 4 & 5 \\ 1 & 1 & 0 \\ -2 & 3 & 4 \end{pmatrix}$$

are multiplied together to give a 3×3 matrix AB in which the element in the rth row and sth column is a_{rs}. [*Hint.* For a_{32}, $r = 3$, $s = 2$.]

5. Matrix Method of Solution of Simultaneous Linear Equations

We now investigate in detail the method of solving simultaneous linear equations mentioned at the beginning of this chapter.

If we apply the rules of multiplication to, say,

$$\begin{pmatrix} 3 & 1 \\ 4 & -1 \end{pmatrix} \begin{pmatrix} x \\ y \end{pmatrix} \quad \text{we get} \quad \begin{pmatrix} 3x+y \\ 4x-y \end{pmatrix}.$$

If the result is $\begin{pmatrix} 5 \\ 2 \end{pmatrix}$, as in the example on p. 164, we have

$$\begin{pmatrix} 3x+y \\ 4x-y \end{pmatrix} = \begin{pmatrix} 5 \\ 2 \end{pmatrix},$$

and applying the rule for the identity (equality) of matrices, we have

$$3x+y = 5, \tag{1}$$

$$4x-y = 2. \tag{2}$$

We have thus shown that writing the pair of equations (1), (2) above as

$$\begin{pmatrix} 3 & -1 \\ 4 & -1 \end{pmatrix} \begin{pmatrix} x \\ y \end{pmatrix} = \begin{pmatrix} 5 \\ 2 \end{pmatrix}$$

is mathematically sound, but it does not at the moment appear helpful.

If, however, we can find a matrix of numbers (not including x and y) by which to multiply both sides of the equation above so that the left hand side becomes $\begin{pmatrix} x \\ y \end{pmatrix}$, our problem is virtually solved.

We start by considering the symmetrical matrix

$$\begin{pmatrix} 1 & 0 \\ 0 & 1 \end{pmatrix},$$

which we call a unit matrix and designate **I**. Let us operate with it on the general 2×2 matrix (which will here be named **A**).

$$\mathbf{IA} = \begin{pmatrix} 1 & 0 \\ 0 & 1 \end{pmatrix} \begin{pmatrix} a_{11} & a_{12} \\ a_{21} & a_{22} \end{pmatrix} = \begin{pmatrix} a_{11} & a_{12} \\ a_{21} & a_{22} \end{pmatrix} = \mathbf{A},$$

on applying the matrix law of multiplication given in Section 4 above. Thus, operating with the unit matrix on another compatible matrix leaves the latter unchanged. We can easily extend the idea to prove that

$$\mathbf{IA} = \mathbf{A} = \mathbf{AI}$$

for any square matrix **A**, provided that **I** is a unit matrix of the same dimensions, e.g. **I** of order 3×3 is

$$\begin{pmatrix} 1 & 0 & 0 \\ 0 & 1 & 0 \\ 0 & 0 & 1 \end{pmatrix}$$

The general definition of **I** is a square matrix whose principal diagonal (top left to bottom right) is a series of 1's and all of whose other elements are zeros.

In algebra, if we wish to multiply a by a number in order to get 1, we can see at once that we need $\dfrac{1}{a}$,

for
$$\frac{1}{a} \times a = 1.$$

This can be written

$$a^{-1}a = 1$$

using the law of indices.

In matrix algebra we accept an analogous notation

$$\mathbf{A}^{-1}\mathbf{A} = \mathbf{I},$$

where $\mathbf{A}^{-1}$ is to be determined when $\mathbf{A}$ is known.

Consider the simultaneous equations in x and y,

$$ax + by = h$$
$$cx + dy = k,$$

which can be written

$$\mathbf{AX} = \mathbf{B}$$

where
$$\mathbf{A} = \begin{pmatrix} a & b \\ c & d \end{pmatrix}, \quad \mathbf{X} = \begin{pmatrix} x \\ y \end{pmatrix}, \quad \mathbf{B} = \begin{pmatrix} h \\ k \end{pmatrix}.$$

We can solve this if we can find a matrix $\mathbf{A}^{-1}$, called the *inverse* of $\mathbf{A}$, such that

$$\mathbf{A}^{-1}\mathbf{AX} = \mathbf{A}^{-1}\mathbf{B}$$
$$\Rightarrow \quad \mathbf{IX} = \mathbf{A}^{-1}\mathbf{B}$$
$$\Rightarrow \quad \mathbf{X} = \mathbf{A}^{-1}\mathbf{B}.$$

Suppose $\mathbf{A}^{-1} = \begin{pmatrix} p & q \\ r & s \end{pmatrix}$, then

$$\mathbf{A}^{-1}\mathbf{A} = \begin{pmatrix} p & q \\ r & s \end{pmatrix}\begin{pmatrix} a & b \\ c & d \end{pmatrix} = \begin{pmatrix} pa+qc & pb+qd \\ ra+sc & rb+sd \end{pmatrix}.$$

Put $pb + qd = 0$ and $ra + sc = 0$, then

$$\frac{p}{d} = -\frac{q}{b} = k_1 \text{ (say) and } \frac{r}{c} = -\frac{s}{a} = k_2 \text{ (say)},$$

where k_1 and k_2 are to be determined.

Then $pa+qc = k_1(ad-bc)$ and $rb+sd = k_2(bc-ad)$.

$$\therefore \mathbf{A}^{-1}\mathbf{A} = \begin{pmatrix} k_1(ad-bc) & 0 \\ 0 & k_2(bc-ad) \end{pmatrix}$$

on substituting for p, q, r, s. This further simplifies to

$$\mathbf{A}^{-1}\mathbf{A} = [ad-bc]\begin{pmatrix} k_1 & 0 \\ 0 & -k_2 \end{pmatrix},$$

where $ad-bc$ is called the *determinant* of $\mathbf{A}$ and may be written Δ for short. (*Note.* Δ is a number, *not* a matrix.)

Finally, putting $k_1 = \dfrac{1}{\Delta}$ and $k_2 = -\dfrac{1}{\Delta}$,

we have
$$\mathbf{A}^{-1}\mathbf{A} = \Delta \begin{pmatrix} \dfrac{1}{\Delta} & 0 \\ 0 & \dfrac{1}{\Delta} \end{pmatrix} = \Delta.\dfrac{1}{\Delta}.I = I,$$

where
$$\frac{p}{d} = -\frac{q}{b} = \frac{1}{\Delta} \text{ and } \frac{r}{c} = -\frac{s}{a} = -\frac{1}{\Delta}$$

$$\therefore \mathbf{A}^{-1} = \frac{1}{\Delta}\begin{pmatrix} d & -b \\ -c & a \end{pmatrix},$$ and the student may find it helpful to memorize this. If $\Delta = 0$, the method fails. [See Ex. 39, No. 4.]

The method can be extended to any number of simultaneous linear equations, provided that the number of unknowns is the same as the number of equations. *It is this extension and its application to computing which lead to the current widespread interest in matrices.*

We shall now solve the example given at the beginning of the chapter and referred to on several occasions.

Example. Solve $3x+y = 5$
$\qquad\qquad 4x-y = 2.$

We have $\mathbf{A} = \begin{pmatrix} 3 & 1 \\ 4 & -1 \end{pmatrix}$, $\mathbf{X} = \begin{pmatrix} x \\ y \end{pmatrix}$, $\mathbf{B} = \begin{pmatrix} 5 \\ 2 \end{pmatrix}$

$\Rightarrow \qquad \mathbf{A}^{-1} = \dfrac{1}{-3-4}\begin{pmatrix} -1 & -1 \\ -4 & 3 \end{pmatrix} = -\dfrac{1}{7}\begin{pmatrix} -1 & -1 \\ -4 & 3 \end{pmatrix}$

so $\qquad \mathbf{AX} = \mathbf{B}$

becomes
$$\mathbf{X} = -\frac{1}{7}\begin{pmatrix} -1 & -1 \\ -4 & 3 \end{pmatrix}\begin{pmatrix} 5 \\ 2 \end{pmatrix}$$

$$= -\frac{1}{7}\begin{pmatrix} -7 \\ -14 \end{pmatrix} = \begin{pmatrix} 1 \\ 2 \end{pmatrix}$$

$$\therefore \qquad \begin{pmatrix} x \\ y \end{pmatrix} = \begin{pmatrix} 1 \\ 2 \end{pmatrix}$$

$$\Rightarrow \qquad x = 1, y = 2$$

[Check $3.1 + 2 = 5$; $4.1 - 2 = 2$.]

Exercise 39

Solve the following pairs of simultaneous equations, using matrices (Nos. 1–4):

1. $3x + 2y = 6$, $x + 5y = 2$.
2. $3m + 2n = -1$, $6(m - n) = 13$.
3. $x + 2y = 2$, $4y - 3x = 7$.
4. $5x + 4y = 3$, $10x + 8y = 7$. Why does the method fail?
[*Note*: This is an important special case.

$$\Delta = 5.8 - 4.10 = 0 \text{ and } \frac{1}{\Delta} \text{ would be infinite.}$$

In the circumstances $\mathbf{A}^{-1}$ cannot be determined and we say that the matrix $\mathbf{A}$ is *singular*.

The explanation is that there are no *finite* solutions for x and y. If we graph the lines $5x + 4y = 3$, $10x + 8y = 7$, we find that they are parallel. They can be said to meet on the *line at infinity* and the solution of the simultaneous equations would be infinite.]

5. If $[4a^2 - b^2]\mathbf{A} = \begin{pmatrix} 4a & -b \\ -b & a \end{pmatrix}\begin{pmatrix} a & b \\ b & 4a \end{pmatrix}$,
prove that $\mathbf{A} = \mathbf{I}$.

6. Power of a Matrix

We have considered the multiplication of matrices such as $\mathbf{AB}$ where $\mathbf{A}$ and $\mathbf{B}$ are compatible. Let us now consider $\mathbf{AA} = \mathbf{A}^2$ and $\mathbf{AAA} = \mathbf{AA}^2 = \mathbf{A}^2\mathbf{A} = \mathbf{A}^3$. For simplicity we restrict our considerations to 2×2 matrices. The theory can be extended to higher orders but it is interesting to note that powers of $\mathbf{A}$ (such as $\mathbf{A}^2$, $\mathbf{A}^3$)

only have a meaning if A is a square matrix. This is easily seen, for $(a_{mn}) \times (a_{mn})$ only has a meaning if $m = n$.

Let
$$\mathbf{A} = \begin{pmatrix} a & b \\ c & d \end{pmatrix}$$

then
$$\mathbf{A}^2 = \begin{pmatrix} a & b \\ c & d \end{pmatrix}\begin{pmatrix} a & b \\ c & d \end{pmatrix} = \begin{pmatrix} a^2+bc & ab+bd \\ ac+cd & bc+d^2 \end{pmatrix}$$

$$= \begin{pmatrix} a^2+bc & b[a+d] \\ c[a+d] & bc+d^2 \end{pmatrix}$$

$$= \begin{pmatrix} a[a+d]+[bc-ad] & b[a+d] \\ c[a+d] & d[a+d]+[bc-ad] \end{pmatrix}$$

where the artificial insertion of $+ad-ad$ in two places is made to create a common factor $[a+d]$ already present in two elements.

$$\therefore \qquad \mathbf{A}^2 = \begin{pmatrix} a[a+d] & b[a+d] \\ c[a+d] & d[a+d] \end{pmatrix} + \begin{pmatrix} bc-ad & 0 \\ 0 & bc-ad \end{pmatrix}$$

$$= [a+d]\begin{pmatrix} a & b \\ c & d \end{pmatrix} - [ad-bc]\begin{pmatrix} 1 & 0 \\ 0 & 1 \end{pmatrix}$$

$$= [a+d]\,\mathbf{A} - \Delta\mathbf{I},$$

where we must retain **I** here, as each term must have a matrix in it, and where $[a+d]$ is a scalar number and Δ has the definition given on page 179. Hence we have the *characteristic equation* of the matrix **A**, i.e.

$$\mathbf{A}^2 - [a+d]\mathbf{A} + \Delta\mathbf{I} = \phi.$$

This is very useful for finding higher powers of **A**. Suppose we multiply throughout by **A**, then

$$\mathbf{A}^3 - [a+d]\mathbf{A}^2 + \Delta\mathbf{A} = \phi$$

for $\mathbf{AI} = \mathbf{A}$ and $\mathbf{A}\phi = \phi$.

Hence, using the characteristic equation already found,

$$\mathbf{A}^3 - [a+d]\{[a+d]\mathbf{A} - \Delta\mathbf{I}\} + \Delta\mathbf{A} = \phi$$

$$\Rightarrow \qquad \mathbf{A}^3 = \{[a+d]^2 - \Delta\}\mathbf{A} - [a+d]\Delta\mathbf{I}.$$

Example. If $\mathbf{A} = \begin{pmatrix} 3 & -2 \\ 0 & 1 \end{pmatrix}$, find $\mathbf{A}^3$.

(Method 1) Here $a = 3, b = -2. c = 0, d = 1,$

$\therefore \qquad a+d = 4$ and $\Delta = ad - bc = 3$

$\therefore \qquad \mathbf{A}^3 = (16-3)\mathbf{A} - 4.3\mathbf{I}$

$$= 13\mathbf{A} - 12\mathbf{I}.$$

$$= \begin{pmatrix} 39 & -26 \\ 0 & 13 \end{pmatrix} - \begin{pmatrix} 12 & 0 \\ 0 & 12 \end{pmatrix}$$

$$= \begin{pmatrix} 27 & -26 \\ 0 & 1 \end{pmatrix}$$

(Method 2) $\qquad \mathbf{A}^2 = \begin{pmatrix} 3 & -2 \\ 0 & 1 \end{pmatrix}\begin{pmatrix} 3 & -2 \\ 0 & 1 \end{pmatrix}$

$$= \begin{pmatrix} 9 & -8 \\ 0 & 1 \end{pmatrix}$$

$$\mathbf{A}^3 = \mathbf{A}\mathbf{A}^2 = \begin{pmatrix} 3 & -2 \\ 0 & 1 \end{pmatrix}\begin{pmatrix} 9 & -8 \\ 0 & 1 \end{pmatrix}$$

$$= \begin{pmatrix} 27 & -26 \\ 0 & 1 \end{pmatrix}$$

In this case the direct method is a little shorter but this is not necessarily so, in general.

We now return to the general case of a 2×2 matrix and show that the characteristic equation gives an effective and surprisingly simple method of finding the inverse $\mathbf{A}^{-1}$ of $\mathbf{A}$.

If $\mathbf{A} = \begin{pmatrix} a & b \\ c & d \end{pmatrix}$, we have already found that

$$\Delta\mathbf{I} = [a+d]\mathbf{A} - \mathbf{A}^2$$

Multiplying by $\mathbf{A}^{-1}$ and noting that $\mathbf{A}^{-1}\mathbf{I} = \mathbf{A}^{-1}$, $\mathbf{A}^{-1}\mathbf{A} = \mathbf{I}$, $\mathbf{A}^{-1}\mathbf{A}^2 = \mathbf{A}$, we have

$$\Delta\mathbf{A}^{-1} = [a+d]\mathbf{I} - \mathbf{A}$$

$$= \begin{pmatrix} a+d & 0 \\ 0 & a+d \end{pmatrix} - \begin{pmatrix} a & b \\ c & d \end{pmatrix}$$

$$= \begin{pmatrix} d & -b \\ -c & a \end{pmatrix}$$

i.e. $\qquad \mathbf{A}^{-1} = \dfrac{1}{\Delta}\begin{pmatrix} d & -b \\ -c & a \end{pmatrix}$, if $\Delta \neq 0$,

a result already found on page 179.

Exercise 40

Find the characteristic equations of the matrices in Qns. 1–4:

1. $\begin{pmatrix} 2 & 1 \\ 0 & -1 \end{pmatrix}$
2. $\begin{pmatrix} a & b \\ -b & a \end{pmatrix}$

3. $\begin{pmatrix} 5 & -8 \\ 4 & -5 \end{pmatrix}$
4. $\begin{pmatrix} 2c & d \\ 4c & 2d \end{pmatrix}$

5. Do the inverses of the following matrices $\mathbf{P}$ and $\mathbf{Q}$ exist

$$\mathbf{P} = \begin{pmatrix} 4 & 2 \\ 2 & 1 \end{pmatrix}, \quad \mathbf{Q} = \begin{pmatrix} 4 & 2 \\ 2 & -1 \end{pmatrix} ?$$

If possible, write them down. [*Hint*. Is Δ zero?]

6. If $\mathbf{A} = \begin{pmatrix} a & a \\ b & -b \end{pmatrix}$, find $\mathbf{A}^3$ in terms of a and b.

7. If $\mathbf{B} = \begin{pmatrix} 3 & -1 \\ 2 & 4 \end{pmatrix}$, find $\mathbf{B}^4$ in the form $49\mathbf{C}$, where $\mathbf{C}$ is a matrix.

[*Hint*. $\mathbf{B}^4 = \mathbf{B}^2\mathbf{B}^2$.]

8. Solve the simultaneous equations in x and y:

(a) $2x - y = 5$, $\quad x + 2y = 3$

(b) $ax + by = c$, $\quad bx + 4ay = -2c$,

by a matrix equation of the form $\mathbf{AX} = \mathbf{B} \Rightarrow \mathbf{X} = \mathbf{A}^{-1}\mathbf{B}$ where $\mathbf{A}^{-1}$ is found as in Section 6 above.

7. Application of Matrices to Public Opinion Polls

We complete the chapter with a short consideration of real-life problems. Suppose that an opinion poll is taken of a cross-section of the population in order to find out its views on corporal punishment in schools. It is found that 40% favour it and that the rest are opposed. A little later, the same people are again asked their opinions and it is discovered that $\frac{1}{3}$ of those originally in favour are now opposed, whereas $\frac{1}{4}$ of those originally opposed are now in favour.

We lay out a column matrix of the original distribution:

$$\begin{matrix} \text{For} \\ \text{Against} \end{matrix} \qquad \mathbf{B} = \begin{pmatrix} 40 \\ 60 \end{pmatrix}$$

If we now consider the matrix *operating* on this, we have the situation that we need a 2×2 matrix. Originally it was a unit matrix but after the second opinion poll it has had to be modified.

$$\begin{matrix} & \text{Original Poll} & & & \text{Modified Poll} \\ \text{For} & \mathbf{I} = \begin{pmatrix} 1 & 0 \\ 0 & 1 \end{pmatrix} & \text{becomes} & \mathbf{A} = \begin{pmatrix} 1-\frac{1}{5} & \frac{1}{4} \\ \frac{1}{5} & 1-\frac{1}{4} \end{pmatrix} \\ \text{Against} \end{matrix}$$

The elements of the matrix are in the correct positions, for the top line multipliers operating on the column matrix **B** will give those *for* corporal punishment and the bottom line multipliers will give those *against* it. Thus we have **IB** becomes **AB**.

Original Poll

$$\begin{pmatrix} 1 & 0 \\ 0 & 1 \end{pmatrix}\begin{pmatrix} 40 \\ 60 \end{pmatrix} = \begin{pmatrix} 40 \\ 60 \end{pmatrix}$$

New Poll

$$\begin{pmatrix} \frac{4}{5} & \frac{1}{4} \\ \frac{1}{5} & \frac{3}{4} \end{pmatrix}\begin{pmatrix} 40 \\ 60 \end{pmatrix} = \begin{pmatrix} \frac{4}{5} \times 40 + \frac{1}{4} \times 60 \\ \frac{1}{5} \times 40 + \frac{3}{4} \times 60 \end{pmatrix} = \begin{pmatrix} 47 \\ 53 \end{pmatrix}$$

Instead of 40% for, 60% against, we now have 47% for, 53% against.

If the trend were maintained at this level, after another comparable interval of time we might anticipate the situation as

$$\mathbf{A^2B} = \begin{pmatrix} \frac{4}{5} & \frac{1}{4} \\ \frac{1}{5} & \frac{3}{4} \end{pmatrix}\begin{pmatrix} 47 \\ 53 \end{pmatrix} = \begin{pmatrix} 37 \cdot 6 + 13 \cdot 25 \\ 9 \cdot 4 + 39 \cdot 75 \end{pmatrix}$$

$$= \begin{pmatrix} 50 \cdot 85 \\ 49 \cdot 15 \end{pmatrix}$$

There are now 50·85% in favour of corporal punishment and so public opinion is just biased towards it.

In practice there are usually some people unable to make up their minds about controversial issues. We now consider an example which takes into account this difficulty.

Example. An urban district council wishes to apply for borough status, which will lead to expenses in the provision of a mayor and corporation, on the one hand, but higher status for the community, in the provision of more extensive public services, on the other. An opinion poll finds that 40% are in favour, 50% are opposed and the rest are undecided. Later, it is found that, of those originally in favour $\frac{1}{10}$ are opposed; of those originally opposed, $\frac{1}{8}$ are in favour and $\frac{1}{20}$ are undecided; of those originally undecided, $\frac{1}{4}$ are now in favour. There are no other changes. May the council assume that the local community, even if not enthusiastic, tends to favour the scheme?

We construct a matrix **A** (this time it is 3×3) with which we are going to operate.

	For	Against	Undecided
For	$1-\frac{1}{10}$	$\frac{1}{8}$	$\frac{1}{4}$
Against	$\frac{1}{10}$	$1-\frac{1}{8}-\frac{1}{20}$	0
Undecided	0	$\frac{1}{20}$	$\frac{3}{4}$
	1	1	1

(Note that these columns add up to 1.)

Hence
$$
\mathbf{A} = \begin{pmatrix} \frac{9}{10} & \frac{1}{8} & \frac{1}{4} \\ \frac{1}{10} & \frac{33}{40} & 0 \\ 0 & \frac{1}{20} & \frac{3}{4} \end{pmatrix}
$$

The matrix **B** on which we are operating is given by

For $\qquad$ 40 $\left.\right\}$
Against $\quad$ 50 $\left.\right\} \Rightarrow \mathbf{B} = \begin{pmatrix} 40 \\ 50 \\ 10 \end{pmatrix}$
Undecided $\;$ 10 $\left.\right\}$

$\therefore$
$$
\mathbf{AB} = \begin{pmatrix} \frac{9}{10} & \frac{1}{8} & \frac{1}{4} \\ \frac{1}{10} & \frac{33}{40} & 0 \\ 0 & \frac{1}{20} & \frac{3}{4} \end{pmatrix} \begin{pmatrix} 40 \\ 50 \\ 10 \end{pmatrix}
$$

$$
= \begin{pmatrix} 36+ \;6\cdot25+2\cdot5 \\ 4+41\cdot25+0 \\ 0+ \;2\cdot5 \;+7\cdot5 \end{pmatrix} = \begin{pmatrix} 44\cdot75 \\ 45\cdot25 \\ 10\cdot00 \end{pmatrix}
$$

This means that 44·75% are in favour of the recommendation, 45·25% are opposed to it, and that again 10% of the citizens are undecided. The council ought therefore to abandon the scheme.

Exercise 41

1. In Milchester, a hypothetical small provincial town where the main members of the local council were prejudiced in their views, there was a strong move to instruct the citizens to stop smoking. Before resorting to enforcement, however, the more moderate members persuaded the council to obtain a public opinion poll on the matter. A first test revealed that 70% were opposed to enforcement and the remainder in favour. A second poll, taken a little later, revealed that, of those originally opposed, $\frac{1}{10}$ were now in favour and, of those originally in favour, $\frac{1}{3}$ were now opposed. What was the situation after the second poll?

If a third poll were taken after a similar time interval, assuming that resentment against improper pressure continued to grow at the same rate, what would be the position?

2. In the same small town as above, which would be unlikely to exist in real life, there was great enthusiasm for the construction of swimming baths. A public opinion poll was taken and it was found that 55% were in favour, 20% were undecided and the rest were opposed. Shortly afterwards, a councillor, after a cheerful evening, accidentally let fall the information that the scheme would cause the rates to rise by 25p (i.e. 25 new pence) in the £. There was a public outcry and it was insisted that another poll be taken. It was found that of those in favour, $\frac{4}{5}$ now opposed; of those who were undecided, $\frac{3}{4}$ now opposed; and that of those originally opposed $\frac{1}{20}$ were in favour. There were no other changes. What was the overall position?

3. The manufacturers of three kinds of washing powder were waging their usual advertisement warfare. A first opinion poll was taken to assess the situation and it was found that, of a sample of housewives, 39% favoured Shrinko, 28% liked Rottaway and the rest used Skratch. After various attractions had been offered as free gifts—for example, celluloid ashtrays, plastic socks and wooden screwdrivers —another test of the same housewives was taken. Of those who originally preferred Shrinko, $\frac{1}{5}$ now supported Rottaway and $\frac{1}{10}$ supported Skratch. Of those who had favoured Rottaway, $\frac{1}{8}$ now supported Shrinko and $\frac{1}{6}$ supported Skratch. Of those who had supported Skratch, $\frac{1}{7}$ now supported Shrinko and $\frac{1}{4}$ preferred Rottaway. Find, correct to two decimal places, the percentage of housewives in favour of each product.

CHAPTER 12

PROBABILITY

1. Probability for a Single Event

Although it is not our intention to delve into the *modus operandi* in gambling, it is usually recognised that the simplest introduction to probability is through the medium of pennies, dice and cards.

If we toss a penny we can reasonably expect that it will come to rest either as a head or as a tail. It may, of course, be cross-grained and stick on its edge in a patch of mud or irretrievably be lost down a convenient drain, but we shall ignore such anti-social tendencies. We shall say that the penny will do one of two things—expose a head or a tail—and that this is *certainty*, which we define as 1. It is reasonable to assume that the penny is as likely to land with one face uppermost as with the other. Thus, the theoretical probability of a head is $\frac{1}{2}$ and that of a tail is $\frac{1}{2}$, for $\frac{1}{2}+\frac{1}{2} = 1$, i.e. certainty. Again, for the questing mind, it is of interest to note that, as the two faces of the coin are not identical in design, the probability is very, very slightly skew and it may well be that there is a minute bias towards heads—but it is negligible from the point of view of captains of football teams when tossing for kick-off!

In real life, unless the number of throws is large, the *practical* probability distribution obtained may differ substantially from the *theoretical* result. The author's son, Peter, tossed a new penny fifty times as an illustration. The first set of results was poor:

> Heads 20
> Tails 30.

The experiment was repeated and this time the penny was more co-operative:

> Heads 26
> Tails 24.

The probability, p, of an event happening in an experiment is given by

$$p = \frac{\text{Number of favourable cases}}{\text{Total number of cases}} = \frac{a}{a+b},$$

where a is the number of favourable cases and b is the number of unfavourable cases. Thus, in the above experiments on tossing a penny, the first practical probability of a head was $\frac{20}{50} = 0.4$, compared with the theoretical probability of $\frac{1}{2} = 0.5$. Tabulating the results of both experiments, we have

	Practical Results		Theoretical Result
	1st trial	2nd trial	
Head	0·40	0·52	0·5
Tail	0·60	0·48	0·5
+	1·00	1·00	1·0

It will be observed that if we add each column we obtain 1·00, i.e. certainty. Had we tossed the penny a large number of times, say 1000, we could have expected a result closely akin to the theoretical one.

The probability, p', of an event *not* happening is logically defined as

$$p' = \frac{\text{Number of unfavourable cases}}{\text{Total number of cases}} = \frac{b}{a+b} = \frac{(a+b)-a}{a+b}$$

$$= 1 - \frac{a}{a+b} = 1 - p,$$

which we could have seen by considering

Probability of event happening + Probability of event not happening

$$= \text{Certainty.}$$

In set notation, p' is the complement of p in *certainty*, the universal set ($= 1$).

It should be added that there are events to which this definition does not apply directly.

We now turn our attention to the rolling of a single die having six faces numbered 1 to 6. Suppose it is rolled 300 times to see on how many occasions each face turns upward. Once again the author is indebted to his long-suffering son, who obtained the following cases:

Die face / 300 throws	1	2	3	4	5	6	Total
Actual result	47	45	58	58	53	39	300
Theoretical result	50	50	50	50	50	50	300

The theoretical expectation of a particular face being uppermost is, in this case, $300 \div 6$ (the number of faces), i.e. 50. The number of throws was not large compared with the number of possibilities (six different results) but the poor result for face number six suggested faulty manufacture of the die. This was borne out by subsequent trials and by comparison with other dice. In fact, less biased results would have been obtained by collating information from a group of people each throwing a different die, for then defects of workmanship would have tended to cancel out.

From the results obtained, the practical and theoretical probabilities of throwing 1 are seen to be $\frac{47}{300} \simeq 0 \cdot 157$ and $\frac{50}{300} \simeq 0 \cdot 167$ respectively. Proceeding thus, we have the table:

Face / Probability	1	2	3	4	5	6
Practical	0·157	0·150	0·193	0·193	0·177	0·130
Theoretical	0·167	0·167	0·167	0·167	0·167	0·167

We pursue our investigations, using an ordinary pack of 52 playing cards.

Example. From an ordinary pack of cards, a single card is drawn. What is the probability that it is (*a*) a. heart, (*b*) the king of hearts, (*c*) not a court card, (*d*) red?

(*a*) Of 52 cards, 13 are hearts

$$\therefore \quad p = \frac{13}{52} = \frac{1}{4}$$

(*b*) Of 52 cards, only one (the king of hearts) is acceptable .

$$\therefore \quad p = \frac{1}{52}.$$

(*c*) Of 52 cards, there are 12 court cards (kings, queens, jacks)

$$\therefore \quad p' = \frac{12}{52}, \quad \text{i.e.} \quad p = \frac{40}{52} = \frac{10}{13}.$$

(*d*) Of 52 cards, 26 are red

$$\therefore \quad p = \frac{26}{52} = \frac{1}{2}.$$

We now define the *odds* in favour of an event occurring and the *odds* against it occurring as

$$\text{Odds in favour} = \frac{\text{Number of favourable cases}}{\text{Number of unfavourable cases}} = \frac{a}{b},$$

usually written in the form $a:b$.

$$\text{Odds against} = \frac{\text{Number of unfavourable cases}}{\text{Number of favourable cases}} = \frac{b}{a},$$

usually written in the form $b:a$.

The odds against are therefore the reciprocal of the odds in favour. If we call the odds in favour and odds against o_f and o_a respectively,

$$o_f . o_a = 1$$

which is quite different from the probability relationship

$$p + p' = 1$$

If $p = \dfrac{a}{a+b}$ (as before), we have

$$o_f = \dfrac{a}{b} = \dfrac{a}{a+b} \div \dfrac{b}{a+b}$$

$$= \dfrac{p}{p'}$$

Example. A player draws an ace from a pack of 52 cards. The ace is not returned to the pack. Another player now draws from the pack. What are now the odds against this card also being an ace?

Originally, of 52 cards, 4 were aces.

After the first draw there are 51 cards, of which 3 are aces.

$$\therefore \quad p = \dfrac{3}{51} = \dfrac{1}{17}$$

$\therefore$ Odds against an ace are $(17-1):1$, i.e. $16:1$.

Note. This problem is not the same as that of determining, *before either player has drawn,* the odds against both the first and second cards being aces. A more elaborate situation, of this type, is dealt with under the heading *compound probability* in this chapter.

Laplace's First Principle is a summary of the definitions in this section.

If an event can succeed in a ways and fail in b ways, all ways being equally likely, then:

> the probability (chance) of success is $a/(a+b)$,
> the probability of failure is $b/(a+b)$.
> the odds in favour are $a:b$.
> the odds against are $b:a$.

Exercise 42

1. A boy has tossed a penny six times and it has come down heads on each occasion. He now tosses it a seventh time. What is the probability that it will come down heads?

2. Roll a die 100 times. Add up the number of times it comes up as 2 or 3. Express this as a *practical* probability, in decimal form. What is the *theoretical* probability? Find the percentage error in the practical probability.

(i.e. $\dfrac{\text{Difference between theoretical and practical probability}}{\text{Theoretical probability}} \times 100$).

3. Jones and Smith play a curious gambling game. Jones has an ordinary pack of cards from which kings, queens, knaves, tens and nines have been removed. Aces are counted as ones. Smith has a single die. Jones draws a card at the same time as Smith rolls his die. If the former has a card with a prime number on it, he wins, unless the latter has a prime number on his die, in which case it is a draw. Conversely the latter wins if he has a prime number on his die when the former does not on his card. If Jones replaces the card he draws and shuffles the pack before he draws again, who is likely to win an evening's play?

4. From an ordinary pack of 52 playing-cards, what is the probability that a single card drawn will
    (a) be a red court card, (a court card is a king, queen or knave),
    (b) have a face value which is a multiple of 3 or 5?
(In this sense, knaves, queens and kings do not have a face value).

5. Two cards are drawn from an ordinary pack of 52 and are placed face up on a table. They are found to be the 2 and 3 of clubs. If they are not replaced in the pack, what is the probability that the next card drawn is (a) a club, or (b) a three, or (c) black?

6. A bag contains six white balls and five red balls of the same size. Three balls are drawn without replacing them and are found to be red. What is the probability that the next ball drawn is white? What are the odds in favour of this happening?
    All the balls are replaced in the bag. What is the smallest number of balls which needs to be drawn to be certain of getting (a) three white balls, (b) three balls of one colour?

2. Exclusive Events (Addition Law of Probability)

Suppose that we wish to find the probability of scoring 7 with a single throw of a pair of dice. The possible successful results are

$$(1, 6), \quad (2, 5), \quad (3, 4), \quad (4, 3), \quad (5, 2), \quad (6, 1),$$

where the first number in each pair refers to the first die and the second number to the second die. We have 6 possible successes, each one of which *excludes* any of the others occurring at the same time.

Now the total of all the possibilities is $6 \times 6 = 36$, for each die has 6 faces, any of which may appear. The probability of success is, therefore

$$\frac{6}{36} = \frac{1}{6}.$$

This is a particular case of *Laplace's Second Principle*, which may be enunciated as follows:

If an event can occur in various ways, any one of which excludes the others happening at the same time, the probability p of occurrence of the event is the sum of the probabilities $p_1, p_2 \ldots p_n$ of the separate ways which can occur, i.e. $p = p_1 + p_2 + \ldots + p_n$.

Proof. Suppose that the event can occur in n ways of which the probabilities are $\frac{a_1}{t}, \frac{a_2}{t} \ldots \frac{a_n}{t}$ respectively, where $a_1, a_2 \ldots a_n$ and t are all integers. (If they are not so to start with, then we make them into this form, e.g. $\frac{2}{15}$, 0.4, 0.01 can be expressed as $\frac{40}{300}, \frac{120}{300}, \frac{3}{300}$).

Altogether there are t equally likely ways, of which the event succeeds in $a_1, a_2, \ldots a_n$ ways, but as they entirely exclude one another they must all be different occurrences and their total is $a_1 + a_2 + \ldots + a_n$. The probability of the event happening is thus

$$p = \frac{a_1 + a_2 + \ldots + a_n}{t} = \frac{a_1}{t} + \frac{a_2}{t} + \ldots + \frac{a_n}{t} = p_1 + p_2 + \ldots + p_n.$$

(Our example in brackets yields $\frac{40 + 120 + 3}{300} = \frac{163}{300}$).

Example. Three dice are rolled together once. What is the probability that the score is at least 16?

The ways in which success can occur are tabulated below:

16	17	18
(6, 5, 5)	(6, 6, 5)	(6, 6, 6)
(5, 6, 5)	(6, 5, 6)	
(5, 5, 6)	(5, 6, 6)	
(6, 6, 4)		
(6, 4, 6)		
(4, 6, 4)		

$$6 \quad + \quad 3 \quad + \quad 1 \quad = 10 \text{ ways}$$

Now the total number of possible cases is $6^3 = 216$

$\therefore$ Probability of success $= \dfrac{10}{216} = \dfrac{5}{108}$.

Exercise 43

1. In a single throw of a pair of dice, what is the probability of obtaining (a) a pair (i.e. the same number on each die), (b) a total of 8?

2. Three dice are rolled together. What are the odds against the total being 7? What is the probability of having one-two-three? (i.e. one die coming up 1, one die coming up 2, and one coming up 3).

3. A single die is rolled twice. What is the probability that the sum of the scores is less than 10?

4. Three pennies are tossed together. What are the odds against two tails and one head?

5. Twelve people sit at a round table, two of them being Mary and Harry. What are the odds against Harry sitting next to Mary, assuming there is no particular reason why he should do so?

3. Independent Events (Multiplication Law of Probability)

Events are independent when the occurrence of any one of them is entirely without effect on the others.

Suppose that, in order to win a challenge, it is necessary both to get a six with a single roll of a die and to get a head with a single toss of a penny. Six results are possible with the die and two with the penny:

$$(1, h), (2, h), (3, h), (4, h), (5, h), (6, h)$$
$$(1, t), (2, t), (3, t), (4, t), (5, t), (6, t)$$

where h, t stand for head, tail respectively. Altogether there are 6×2, i.e. 12, possible results, of which only one is successful. The probability of success is therefore $\dfrac{1}{12}$. We observe that this is the same as $\dfrac{1}{6} \times \dfrac{1}{2}$, i.e. the product of the separate probabilities.

Let us now consider another example to see if this is likely to be true in general.

Example. A certain oriental potentate who possessed a mildly sadistical sense of humour used to give a condemned prisoner a chance of escaping the chopping block. The prisoner had to pass through one of three doors, A, B, C. Entry through two of them was harmless but through the third was disastrous. Having been successful in the first round, the prisoner then had to pass through one of a further set of doors, D, E, F. In this case, one was innocuous and the other two led to fatal results. After this, if the prisoner was still alive, he was freed. "Look how generous the terms are," declared the mighty prince, "as three of the doors out of six are safe you have an even chance of living." But was this true?

Consider the pairs of doors through which the prisoner could go, one being selected from the first group of three (A, B, C) and the other from the second group (D, E, F). We have the ordered pairs:

$$(A, D), (A, E), (A, F)$$
$$(B, D), (B, E), (B, F)$$
$$(C, D), (C, E), (C, F).$$

Let us suppose A, B are safe in the first group and D is safe in the second group, then the only safe routes are the ones underlined, i.e. of 9 possibilities, 2 are correct. The probability of success is therefore $\frac{2}{9}$ and the odds are not evens, as the potentate quite misleadingly suggested, but 7:2 against survival.

Let us now compare this with the probabilities of success in each event, say p_1 for doors A, B, C and p_2 for doors D, E, F. Then

$$p_1 = \frac{2}{3}, \quad p_2 = \frac{1}{3} \quad \text{and} \quad p_1 p_2 = \frac{2}{3} \times \frac{1}{3} = \frac{2}{9}$$

which agrees with the result obtained.[17] It is easy, therefore, to imply the general result:

The probability of a compound event p, made up of independent events occurring, is the product of the probabilities $p_1, p_2, \ldots p_n$ of the independent events, i.e.

$$p = p_1 p_2 \cdots p_n$$

[17] The prisoner's problem could equally easily be laid out as a circuit (see Ch. 10) as shown. The solution is then easily seen.

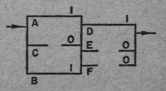

Example. A player rolls a pair of dice once. He wins if *both* dice come to rest with prime numbers on their top faces. What are the odds against him winning?

For each die the prime numbers are 1, 2, 3, 5.

Hence
$$p_1 = \frac{4}{6}, p_2 = \frac{4}{6} \qquad p = p_1 p_2 = \frac{4}{6} \times \frac{4}{6} = \frac{4}{9}$$

$\therefore$ Odds against winning are 5:4.

Exercise 44

1. A die is rolled twice. What is the probability of not getting a 5 on either occasion?

Deduce the probability of getting at least one 5.

[*Hint.* The probability of a non-five is $\frac{5}{6}$ on each occasion.]

2. Two dice are rolled together once. What is the probability that both of them will come up with face values of more than 4?

3. John challenges Henry in the following way. John says "There are six faces on a die and so, if you roll it 3 times you have exactly an even chance of getting a six. I will bet you even money thus— that, if you get a six once (or more) in three throws, I pay you but, if you do not, you pay me." Who is more likely to win? What are the odds in his favour?

[*Hint.* John's argument is false. The event of throwing a six does *not* exclude the event of throwing a six next time, so the events are not exclusive. Hence they cannot be added. As in No. 1 above, we consider the probability of a non-six, i.e. $\frac{5}{6}$. In three throws, this indicates the probability of a non-six every time as $\left(\frac{5}{6}\right)^3$, i.e. the probability of at least one six is $1 - \left(\frac{5}{6}\right)^3$. The reader is left to complete the question.]

4. Interdependent Events

The study of these events is really an extension of the last section, for the reasoning used in Section 3 holds good for the conditions enunciated below.

Suppose that events A and B have probabilities p_1 and p_2 respectively, where p_2 is assessed on the assumption that A has already occurred, then the probability that both A and B occur is $p_1 p_2$.

The idea can be extended to three or more cases.

Example. A small school has two classes, one with 25 boys and 5 girls and the other with 18 boys and 9 girls. One of the classes is chosen at random and an unknown child is called out. What is the probability that the child is a girl?

Would the probability have been the same if all the children had been in one class?

As either class is equally likely, the probability of choosing the first is $\frac{1}{2}$. The probability of a girl is $\frac{5}{30}$, in this class. The probability of having a girl in this class is therefore $\frac{1}{2} \cdot \frac{5}{30}$.

Similarly, the probability of having a girl from the second class is $\frac{1}{2} \cdot \frac{9}{27}$.

The total probability is thus

$$\frac{1}{2} \cdot \frac{5}{30} + \frac{1}{2} \cdot \frac{9}{27} = \frac{1}{12} + \frac{1}{6} = \frac{3}{12} = \frac{1}{4}.$$

If the classes had been combined together, there would have been 43 boys and 14 girls, so the probability of having a girl would have been $\frac{14}{57}$. Although this is not greatly different from the previous result, it certainly is not the same. The explanation is that first choosing a random class biases the result. Had we just asked for a child to walk out of the school, we would have had the second result $\frac{14}{57}$, for any child was equally likely, even if not all had been sitting in the same room.

Exercise 45

1. Two cards are drawn from an ordinary pack. What is the probability that they are (*a*) the ace and king of hearts, (*b*) both diamonds? [*Hint.* In (*a*) the probability of getting a correct card drawn first

time is $\frac{2}{52}$, for either the ace or the king of hearts will do; the prob-

ability of then drawing the correct second card is $\frac{1}{51}$, for having

chosen one of the two possible correct cards, the other is automatically determined out of the remaining 51 cards.]

2. Three cards are drawn from an ordinary pack. What are the odds against them being the king, queen and jack of clubs?

3. From a bag containing 4 red balls and 3 white balls of the same size, two balls are drawn together. What is the probability that there will be one of each colour?

4. Three balls are drawn together from a bag containing 2 red balls, 3 white balls and 4 green balls. Show that the odds against getting one of each colour are 5:2.

5. During a special sale a very large number of postcards is offered at a low price. There are only 3 kinds of card and there is about the same number of each but they have become muddled. A customer must, at the special price, take the first ones which turn up. If he buys five, what is the probability that he has at least one of each kind? [*Hint*. The problem extends the reasoning of the kind used in Ex. 44, No. 3. If T_1 is the first type, the probability of having it at least once is $1-(\frac{2}{3})^5$, and the reader is left to show that this is so. The probability, therefore, of having at least one of T_1, T_2, T_3, the three types, is $\{1-(\frac{2}{3})^5\}^3$. The reader is left to complete this, the answer probably being best expressed in this case as a decimal.]

6. Four prize winners receive their awards from a V.I.P. who does not know them and who forgets to ask their names. What is the probability that all of them will receive the wrong prizes? [*Hint*. The first prize can be given wrongly in 3 ways; the second then wrongly in 3 ways (note: 3, *not* 2); the third and fourth prizes can then only be given in 1 way each, if they are to be wrongly distributed.]

7. The first 100 natural numbers are written down. A friend is asked to cross out any two of them. What are the odds in favour of the sum of the two numbers being odd?

8. There are 29 children at a party, 17 of whom are boys. One boy goes home early. What is the probability that the next two children to leave will be a boy and a girl?

9. The probability that Smith speaks the truth is p_1 and the probability that Brown does is p_2. A certain event may have occurred. (*a*) If Smith and Brown both assert that it has occurred, find the odds

in favour of the event. (*b*) If Smith asserts and Brown denies the occurrence, find the probability that it occurred.

[*Hint.* (*a*) If the event is true, the probability that they both say so is $p_1 p_2$. If it is not true, the probability that they both say so is $(1-p_1)(1-p_2)$, so the odds in favour are $p_1 p_2 : (1-p_1)(1-p_2)$.]

5. Binomial Distribution

Interesting as the foregoing sections are, they have a limited application mainly in the field of gambling. We now see whether we can extend probability theory to cover more constructive practical considerations.

A young married couple hopes to have a family of four children. What is the expectation of having a certain number of boys and a certain number of girls? If the probability of a boy is b and of a girl is g, then the situation is $b+g$ for each birth,[18] i.e. $(b+g)^4$ summarises the problem (for there are 4 births).

We could multiply $(b+g)$ by itself repeatedly but this would be very tedious, so we utilise the Binomial theorem. By actual multiplication we see that

$(b+g)^0$ 1   (by convention)
$(b+g)^1$ $1b+1g$
$(b+g)^2$ $1b^2+2bg+1g^2$
$(b+g)^3$ $1b^3+3b^2g+3bg^2+1g^3$, and so on.

Now we observe that the coefficients can be extended indefinitely:

$$
\begin{array}{ccccccccccc}
 & & & & & 1 & & & & & \\
 & & & & 1 & & 1 & & & & \\
 & & & 1 & & 2 & & 1 & & & \\
 & & 1 & & 3 & & 3 & & 1 & & \\
 & 1 & & 4 & & 6 & & 4 & & 1 & \\
1 & & 5 & & 10 & & 10 & & 5 & & 1
\end{array}
$$

The diagram is called *Pascal's Mystic Triangle.*

---

[18] Note that $b+g=1$ for at each birth, either a boy or a girl is born. Also $g=b'$, but it is more useful to proceed as in the text.

The corresponding powers of b and g are easily seen; for example, in $(b+g)^5$ we start with b^5 and then have b^4g and so on. Thus $(b+g)^5 = Ab^5 + Bb^4g + Cb^3g^2 + Db^2g^3 + Ebg^4 + Fg^5$, where the A, B, C, D, E, F = 1, 5, 10, 10, 5, 1 respectively, from Pascal's triangle.

The general formula for the Binomial expansion is given here for interest but need not be remembered for work in this book:

$$(x+y)^n = x^n + \frac{n}{1!}x^{n-1}y + \frac{n(n-1)}{2!}x^{n-2}y^2 + \ldots$$

$$+ \frac{n(n-1)\ldots(n-r+1)}{r!} x^{n-r}y^r + \ldots + y^n,$$

where $r!$ means $1 \times 2 \times 3 \ldots \times r$, and n is a positive integer.

We can now decide on the probability of certain numbers of children of each sex in our problem. Assume $b = \frac{1}{2}$, $g = \frac{1}{2}$, i.e. each sex is equally likely in any particular birth, then the probability of 2 boys and 2 girls in a family of 4 children is

$$6b^2g^2 = 6\left(\frac{1}{2}\right)^2\left(\frac{1}{2}\right)^2 = \frac{6}{16} = \frac{3}{8}.$$

Similarly the probability of 1 boy and 3 girls is

$$4bg^3 = 4\left(\frac{1}{2}\right)\left(\frac{1}{2}\right)^3 = \frac{4}{16} = \frac{1}{4},$$

and the probability of 4 boys is

$$b^4 = \left(\frac{1}{2}\right)^4 = \frac{1}{16}.$$

One should bear in mind that this calculation presupposes that there is no genetic trait in the family which might predispose one to assume a biased result. If it were known, say, that in past generations there had been 5 girls to every boy, it might be more reasonable to expect that the probability of 2 boys and 2 girls in a family of 4 would be $6b^2g^2$, where $b = \frac{1}{6}$, $g = \frac{5}{6}$, i.e.

$$6\left(\frac{1}{6}\right)^2\left(\frac{5}{6}\right)^2 = \frac{25}{216}.$$

6. Permutations and Combinations

(a) *Permutations.* The number of ways in which n things can be arranged in order when taken r at a time is called *the number of permutations of n things r at a time* and is denoted by ^nP_r.

To find the value of ^nP_r is the same as to find the number of ways of filling r places in order, from n things. The first place can be filled in n ways and the second can then be filled quite independently in $(n-1)$ ways as one thing has now been used. Altogether, from just these two places we have $n(n-1)$ ways. Proceeding thus, we fill the third place in $(n-2)$ ways and so on until we fill the rth place in exactly $(n-r+1)$ ways, for by this time this is the number of things available. Hence

$$^nP_r = n(n-1)(n-2)\ldots(n-r+1)$$

$$= \frac{n(n-1)(n-2)\ldots(n-r+1)(n-r)(n-r-1)\ldots3.2.1}{(n-r)(n-r-1)\ldots3.2.1},$$

on multiplying top and bottom by $(n-r)(n-r-1)\ldots3.2.1$.

$$\therefore \quad ^nP_r = \frac{n!}{(n-r)!}$$

In particular $^nP_n = \frac{n!}{0!} = \frac{n!}{1} = n!$, for we define $0! = 1$.

(b) *Combinations (Selections).* The number of ways in which r things can be selected from n things is called *the number of combinations of n things r at a time* and is denoted by ^nC_r.

From (a) above the total number of ways of arranging n things r at a time is ^nP_r but we do not now wish to consider the order of the r things among themselves. Hence, we have to divide ^nP_r by the total number of repetitive ways of arranging r things taking all of them at a time, i.e. ^rP_r. Now

$$^rP_r = r(r-1)(r-2)\ldots3.2.1 = r!$$

$$\therefore \quad ^nC_r = \frac{^nP_r}{^rP_r}$$

$$= \frac{n!}{r!(n-r)!}$$

An example should make this clear.

Example. In a competition, four people A, B, C, D are equally considered to be worthy of a prize, but only three prizes are available. In how many ways can the prizewinners be chosen (all prizes being the same)?

This is clearly the problem of finding $^4C_3$.

Let us first, however, look at $^4P_3$, the ways of arranging the prizewinners.

ABC	ABD	ACD	BCD
BAC	ADB	ADC	BDC
BCA	BDA	CAD	CBD
CAB	BAD	CDA	CDB
CBA	DBA	DAC	DCB
ACB	DAB	DCA	DBC

These are all different orders ($^4P_3 = 4.3.2 = 24$) and if the prizes were all different this set (the members of $^4P_3$) would be needed. If, however, the prizes are identical, there are only four different results

$$ABC \qquad ABD \qquad ACD \qquad BCD,$$

for $ABC = BCA = CAB = BAC = ACB = CBA$, and similarly for the other groups.

This agrees with $^4C_3$, which is given by

$$^4C_3 = \frac{4!}{3!(4-3)!} = \frac{4!}{3!} = 4 \quad (\text{for } 1! = 1).$$

Many probabilities can be found as a ratio of two permutations *or* as a ratio of two combinations, but it must not be concluded from this statement that the number of such problems is necessarily large compared with the set of problems as a whole.

It will be observed that we have obtained, in ^nC_r, the Binomial coefficient of $x^{n-r}y^r$ in the expansion of $(x+y)^n$.

Example. A warship fires at a line target (i.e. one of negligible thickness compared with the distance of firing). The probability that a single shot falls short is a. What is the probability that in a salvo of n shots, r fall short? If $a = \dfrac{2}{3}$, what is the probability?

The probability of a short for any shot is a; therefore the probability of an over (i.e. a shot which goes too far) is $1-a$. We therefore require the *term* in $a^r(1-a)^{n-r}$ in the Binomial expansion of $\{a+(1-a)\}^n$. Hence the required probability is

$$^nC_r a^r(1-a)^{n-r}$$

$$= \frac{n!}{r!(n-r)!}a^r(1-a)^{n-r}.$$

Suppose, in a salvo of 8 shots, the aim is rather too short. The probability of 6 shorts is, if $a = \frac{2}{3}$ (as given above),

$$\frac{8!}{6!2!}\left(\frac{2}{3}\right)^6\left(\frac{1}{3}\right)^2 = \frac{8.7}{2}\cdot\frac{64}{729}\cdot\frac{1}{9}$$

$$= \frac{1792}{6561}.$$

Exercise 46

1. A penny is tossed 5 times in succession. What is the probability of having two heads and three tails?

Is this the same as tossing 5 pennies once, at the same time?

2. The probability that a television tube is manufactured up to standard is 0·8. A crate containing 6 tubes is tested. What is the probability that (a) all the tubes are of the required standard, (b) 2 tubes are defective? (Give the results to 3 significant figures.)

3. Assuming that a home win, an away win and a draw are equally likely in any football match considered, what are the odds against getting eight draws out of eight results forecast on a football coupon?

4. *De Méré's problem.* What is the probability that, in 24 throws of a pair of dice, double-six will come up at least once? Would it be a winning or a losing bet to back its success with even money?

5. *Count Buffon's Needle.* A series of parallel lines is drawn at *a* inches apart on a large white plain sheet of flat drawing paper. A needle *l* inches long is thrown on to the paper without trying to bias the result, then the probability *p* that the needle lands on a line is given by

$$p = \frac{2l}{\pi a},$$ (proved below)

provided that $l \leqslant a$. Suppose that in N throws, n are successful, then the practical probability is n/N and if N is large this will be reasonably accurate, so

$$\frac{n}{N} = \frac{2l}{\pi a}$$

giving $$\pi = \frac{2lN}{an}$$ (1)

Make such a piece of apparatus. Carry out the experiment 500 times and calculate, from the above formula (1), the value of π.

[*Note.* The formula $p = 2l/\pi a$ is easily proved by reference to the calculus, but those to whom the subject is anathema may omit this bit.

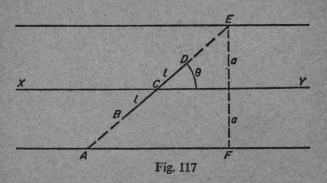

Fig. 117

The needle contacts the line XY if it lies on any part BCD of AE, where θ is the angle the needle makes with XY, and $BC = CD = l$.

Now

$$p_\theta = \frac{BD}{AE} = \frac{2l}{EF \operatorname{cosec} \theta} = \frac{l}{a} \sin \theta,$$

where p_θ is the probability that the needle, in direction θ, contacts a line.

Summing for all cases $0 \leqslant \theta < \pi$

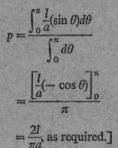

$$p = \frac{\int_0^\pi \frac{l}{a}(\sin \theta)d\theta}{\int_0^\pi d\theta}$$

$$= \frac{\left[\frac{l}{a}(-\cos \theta)\right]_0^\pi}{\pi}$$

$$= \frac{2l}{\pi a}, \text{ as required.]}$$

6. Four dice are rolled together. What is the probability of (a) exactly two 3's, (b) at least two 3's, appearing?

7. A bag contains five 2p pieces and a 5p piece. Another bag contains six 2p pieces. Five coins are taken at random from the first bag and are placed in the second, which is then shaken up. Five coins are then taken from the second bag and are placed in the first. What is the probability that the 5p piece is now in the first bag? [*Hint.* This is an example of interdependent events.]

7. Sets and Probability

We finish with some notes indicating the relationship between sets and probability. Our investigations could, throughout the chapter, have been carried out in set notation, but in many ways it seemed easier firstly to introduce the private student to probability via the medium of cards, coins and dice, with which he is familiar.

We introduce the concept of a *sample space*, U, for a finite set of possible outcomes to a series of events. Suppose we consider Section 3 (p. 194) again, wherein we rolled a die and tossed a penny. Instead of laying out the possible results as ordered pairs, we could use H for head and T for tail and add suffixes to indicate which face of the die turned up, thus H_4 would mean that in this case we had a head and a 4 [i.e. (4, h) in the notation of Section 3]. Hence

$$U = \{H_1, H_2, H_3, H_4, H_5, H_6, T_1, T_2, T_3, T_4, T_5, T_6\},$$

and $n(U)$, meaning the number of members of U, is 12.

If p is an assertion about the sample space, we define P as the truth set of p and state that this is the subset of U for which p is true, then the probability of p, written $\Pr(p)$, is given by[19],

$$\Pr(p) = \frac{n(P)}{n(U)}$$

i.e. *the number of members of subset P divided by the number of members of the sample space.*

Example. Suppose we toss a penny and roll a die. We require the probability of getting a head and an even number.

We have p = "the coin comes up heads and the die comes up even". Using the value of $U = \{H_1, H_2 \ldots H_6, T_1, \ldots T_5, T_6\}$ given above,

$$\Pr(p) = \frac{n(P)}{n(U)} = \frac{3}{12} = \frac{1}{4},$$

for $P = \{H_2, H_4, H_6\} \Rightarrow n(P) = 3$.

We now extend the idea to two assertions p and q, whose truth sets are P and Q respectively. We investigate $n(P) + n(Q)$, where *plus* is used in the ordinary sense of addition as in arithmetic. There are two possibilities:

(i) P and Q have some common members, (ii) P and Q have no common members. (Figs. 118A and 118B.)

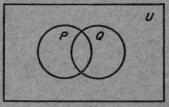

Fig. 118A

[19] The definition is in accordance with the relationship between sets (Chapter 8) and Logic (Chapter 9).

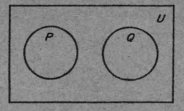

Fig. 118B

(i) In the case of Fig. 118A the members common to P and Q must appear twice in $n(P)+n(Q)$. Now $n(P\cup Q)$ will have members of P not in Q, members of Q not in P and (once) members common to P and Q. Also, $n(P\cap Q)$ will have members common to P and Q.

$$\therefore \quad n(P)+n(Q) = n(P\cup Q)+n(P\cap Q).$$

Dividing by $n(U)$, we have

$$\frac{n(P)}{n(U)}+\frac{n(Q)}{n(U)} = \frac{n(P\cup Q)}{n(U)}+\frac{n(P\cap Q)}{n(U)}$$

$$\Rightarrow \quad \text{Pr}(p)+\text{Pr}(q) = \text{Pr}(p \vee q)+\text{Pr}(p \wedge q).$$

An example should make this clear.

Example. Suppose we take two simple sets $P = \{2, 3, 5, 7, 9\}$ and $Q = \{1, 3, 4, 7\}$, then $n(P) = 5$ and $n(Q) = 4$

$$\therefore \quad n(P)+n(Q) = 5+4 = 9.$$

Also $n(P \cup Q) = n\{1, 2, 3, 4, 5, 7, 9\} = 7$

and $n(P \cap Q) = n\{3, 7\} = 2$

$$\therefore \quad n(P\cup Q)+n(P\cap Q) = 7+2 = 9.$$

Hence $n(P)+n(Q) = n(P\cup Q)+n(P\cap Q)$.

The student will observe that this part of the theorem is true in simple set theory and is only applied to probability in the subsequent stages of reasoning.

(ii) We now discuss the case of Fig. 118B, wherein P and Q have no common members,

$$\text{i.e.} \quad P\cap Q = \phi \Rightarrow n(P \cap Q) = 0 \Rightarrow \text{Pr}(p \wedge q) = 0.$$

The probability equation above then reads

$$\Pr(p \lor q) = \Pr(p) + \Pr(q).$$

It can be extended:

$$\Pr(p \lor q \lor r \ldots) = \Pr(p) + \Pr(q) + \Pr(r) + \ldots$$

This is the equation for *exclusive events* (Ch. 12, Section 2), for it can be read "The probability of an event, which can occur in various ways each excluding any other happening at the same time, is the sum of the separate probabilities. It must be borne in mind that $p \land q = 0$ implies exclusiveness, for it states that assertion p *and* assertion q cannot be true together.

We have, above, constructed logically the rule of addition when events are exclusive. We now turn our attention to events which are not exclusive and see whether we can evolve the product rule of Ch. 12, Sections 3 and 4.

Let us suppose two events take place, the sample space being U. We assume that p actually occurs, e.g. $p =$ "a heart is drawn". The probability of q, say, "the ace of hearts is drawn" is now conditional on p. It is called the conditional probability of q given p and it is written $\Pr(q \mid p)$. A moment's thought will show us that, in thinking of the second event here, we have replaced our original sample space U by the new sample space P (which is the truth set of p and is a sub-space of U). Hence we need to know the number of members of P which are in Q, i.e. $P \cap Q$.

$$\therefore \quad \Pr(q \mid p) = \frac{n(P \cap Q)}{n(P)}$$

$$= \frac{n(P \cap Q)}{n(U)} \div \frac{n(P)}{n(U)}$$

$$= \frac{\Pr(p \land q)}{\Pr(p)}$$

If finally the two events p and q are *independent* (for example, instead of $q =$ "the ace of hearts is drawn" given above, we might have $q =$ "a penny comes down tails", which is quite independent of what we may have just done with a pack of cards), then the truth of p does not influence the truth of q. It follows that

$$\Pr(q \mid p) = \Pr(q).$$

Substituting in the equation immediately above, we have

$$\Pr(q) = \frac{\Pr(p \land q)}{\Pr(p)}$$

$$\Rightarrow \quad \Pr(p \land q) = \Pr(p) . \Pr(q).$$

It can be extended: $\Pr(p \land q \land r \ldots) = \Pr(p) . \Pr(q) . \Pr(r) \ldots$
This can be read "The probability of a number of independent events is the product of their separate probabilities".

Example. Three dice are rolled together. What is the probability that they will all come up with face values of 2 or 1?

Here $U = \{1, 2, 3, 4, 5, 6\}$, $P = \{1, 2\}$, $Q = \{1, 2\}$, $R = \{1, 2\}$.

$$\therefore \quad \Pr(p) = \frac{n(P)}{n(U)} = \frac{2}{6} = \frac{1}{3}; \quad \text{similarly } \Pr(q) = \Pr(r) = \frac{1}{3}$$

$$\therefore \quad \Pr(p \land q \land r) = \frac{1}{3} . \frac{1}{3} . \frac{1}{3} = \frac{1}{27}.$$

Exercise 47

1. In a small school of 80 children, 6 are named Jones and 5 are named Smith. What is the probability that, on the same day, the first child to arrive at school one morning is named Smith and the last to leave in the same evening is named Jones?
2. Three dice are rolled together. What is the probability that the score made is greater than 14?
3. Ten beads are threaded on a circular wire, 5 beads being white and 5 red. What is the probability that they are alternately white and red around the wire?
4. Work as many of the questions of Exercises 42–46 as possible, using set notation.

ANSWERS

Chapter 1

2. (a) 13, (b) 20, (c) 3, (d) 10, (e) 42, (f) 125, (g) 35, (h) 113, (i) 20, (j) 132, (k) 350, (l) 203.
3. (Denary answers) (a) 9, (b) 12, (c) 3, (d) 6, (e) 26, (f) 53, (g) 23, (h) 45, (i) 12, (j) 56, (k) 138, (l) 75.

1. (a) 3, (b) 3, $-2, 1\frac{1}{3}, -\sqrt{16}, 4\cdot3, 0$, (c) 3, $-2, \sqrt{5}, 1\frac{1}{3}, -\sqrt{16}, 4\cdot3, 2+\sqrt{7}, 0$, (d) $\sqrt{-2}$.
3. 70 and 99; 169 and 239; 1·4142.　　4. 2·44 and 2·45.
5. $\dfrac{11}{90}, \dfrac{2}{15}$.　　6. 2·65; 3·16; 4·12.

Chapter 2

1. (a) $-4i$ (imaginary), (b) -1 (real), (c) $-128i$ (imaginary), (d) $-6i$ (imaginary), (e) $\frac{1}{2}$ (real).
3. (a) $6i$, (b) $2i$, (c) $-12i$, (d) $-4i$, (e) $8i$, (g) 6.
4. $\pm\dfrac{3}{2}; \pm\dfrac{7i}{2}$.

1. Imaginary; real; imaginary; real; $a+ib$; $a+ib$; imaginary; $a+ib$; real; imaginary.
2. $i; -1; -4i; -3; 4-5i; 7+3i; -5i; 2-2i; 2; i$.
3. (a) $17+8i$, (b) 5, (c) $-5-10i$.
4. (a) $3+4i$, (b) $10-5i$, (c) $\frac{1}{5}(2-11i)$, (d) $-i$.
5. (a) $\frac{1}{10}(7+9i)$, (b) $1+i$.　　6. 13.

Exercise 5
2. (a) $(5, -5)$, (b) $(2, 0)$, (c) $(-2, 1)$, (d) $(-1, 0)$, (e) $(2ab, b^2-a^2)$,
(f) $(0, a^2+b^2)$, (g) $(y, -x)$, (h) $(14, 22)$, (i) $(\frac{1}{10}, -\frac{4}{5})$, (j) $(-\frac{2}{13}, -\frac{3}{13})$.
3. $2\sqrt{26}-6\sqrt{2} \fallingdotseq 1.71$.

Chapter 3

Exercise 6
1. (a) $(5, 5\sqrt{3})$, (b) $(4\sqrt{2}, 4\sqrt{2})$, (c) $(3, 0)$, (d) $(0, 5)$, (e) $(\frac{1}{2}\sqrt{3}, \frac{1}{2})$,
(f) $(-1, \sqrt{3})$, (g) $(-3\sqrt{2}, -3\sqrt{2})$.
2. (a) $(5, 53°)$, (b) $(5, 307°)$, (c) $(5, 127°)$, (d) $(5, 233°)$ [to the nearest degree].
3. (a) $(-2, 2)$, (b) $(2\sqrt{2}, 135°)$.
4. (i) $\sqrt{17}+\sqrt{10}-\sqrt{29} \fallingdotseq 1.90$, (ii) $(4.12, 76°)$, $(5.39, 112°)$, (iii) $36°$.
5. 3.61 ; $106°$.

Exercise 7
1. (i) $(-2, 5)$, (ii) $(5, 1)$, (iii) $(0, 9)$, (iv) $(-1, 0)$, (v) $(-2, -12)$, (vi) $(0, 6)$.
3. $(5, 37°)$.   4. $2\sqrt{2} \fallingdotseq 2.83$; 315.   5. $\mathbf{p+q}$.   6. $\mathbf{b-a}$.
7. $\mathbf{a+b, b-a}$; $\mathbf{a-b}$.   8. $-\mathbf{b-c-d}$.
9. $2\mathbf{a+b}$; $2\mathbf{a+b+c}$; $\mathbf{a+b+c-d}$.   10. $\frac{1}{2}(\mathbf{a+b})$.

Exercise 8
1. (a) 90 kn., (b) 150 kn., (c) 3 hr. 44 min.   2.   117 knots; $S23°E$.
3. $S25°W$.   4. $N34°E$; 28·8 nautical miles.
5.   Outward journey: $S54°W$, 184 km/h, 65 min.; Return journey: $N35°E$, 128 km/h, 94 min.

Exercise 9
1. $\mathbf{a}+k\mathbf{r}$.   2. $\mathbf{b-a}$; $(1-k)\mathbf{a}+k\mathbf{b}$.

Chapter 4

Exercise 10
1. (a) 123, (b) 125, (c) 0.   2.   (a) 2152, (b) 30343, (c) 1005143.
3.   (a) 131, (b) 23, (c) 125.   4.   (a) 718, (b) 2739, (c) 7777.
5.   (a) 10503, (b) 114433, (c) 1330154.   7.   (a) 9, (b) 8.

Exercise 11
1.   (a) 1100, (b) 10111, (c) 1001000, (d) 10111111, (e) 1000100011,
(f) 101111101110, (g) 100101100001.
2.   (a) 7, (b) 10, (c) 15, (d) 26, (e) 45, (f) 183, (g) 228, (h) 741.
3.   (a) 1100, (b) 11000, (c) 10100, (d) 1011101, (e) 101111,
(f) 110010100.   4.   1100.   5.   (a) 111, (b) 11, (c) 0, (d) 1110.

Exercise 12   *page* 68
1. (*a*) 100011, (*b*) 100001, (*c*) 1001110, (*d*) 100011110, (*e*) 1101001,
    (*f*) 10011011010.
2. (*a*) 110, (*b*) 111, (*c*) 10101.   3.   (*a*) 11, (*b*) 101, (*c*) 11.
4. $x = 10$, $y = 1$.   5.   (*a*) 43, (*b*) 918, (*c*) 27805.
6. (*a*) 1011000, (*b*) 110, (*c*) 10111.   7.   (*a*) 100011, (*b*) 110110.
8. (*a*) 110, (*b*) 10011, (*c*) 100101.   9.   10011.

Exercise 13   *page* 70
1. (*a*) $\frac{3}{4}$; 0·75, (*b*) $\frac{5}{16}$; 0·3125, (*c*) $\frac{19}{32}$; 0·59375, (*d*) $5\frac{1}{4}$; 5·25,

    (*e*) $6\frac{13}{16}$; 6·8125, (*f*) $\frac{1}{2}$; 0·5 [*Note.* $0·0\dot{1}_2 = 0·1_2 = \frac{1}{2}$.]

    (*g*) $9\frac{2}{3}$; $9·\dot{6}$ [*Note.* $0·\dot{1}\dot{0}_2 = \frac{10}{11}$ (scale of 2) $= \frac{2}{3}$ (scale of 10.)]

    (*h*) $\frac{1}{6}$; $0·1\dot{6}$, (*i*) $\frac{3}{7}$; $0·\dot{4}2857\dot{1}$, (*j*) $\frac{3}{5}$; 0·6.

Exercise 14   *page* 73
1. (*a*) 0·011, (*b*) 0·1011, (*c*) $0·\dot{0}1\dot{0}$, (*d*) $0·\dot{1}0001\dot{1}$,
    (*e*) $0·\dot{1}11\dot{0}$, (*f*) $0·10\dot{0}\dot{1}$.
2. (*a*) 0·101, (*b*) 0·0111, (*c*) 0·00101, (*d*) 0·001011.
3. (*a*) $0·\dot{1}10\dot{0}$, (*b*) (*b*) $0·\dot{1}00\dot{1}$, (*c*) $0·1\dot{1}10\dot{0}$, (*d*) $0·11\dot{0}11\dot{0}$, (*e*) $0·\dot{1}\dot{0}$,
    (*f*) $0·\dot{1}11100\dot{0}$, (*g*) $0·\dot{1}11\dot{0}$ [The denary form is $\frac{14}{15}$, which is the
    same as Qn. 1 (*e*) in the same exercise].

Exercise 15   *page* 74
1. (*a*) 117, (*b*) 224, (*c*) 734, (*d*) 7516, (*e*) 24745, (*f*) 60552, (*g*) 234064.
2. (*a*) 100110, (*b*) 111100, (*c*) 100010011, (*d*) 101111110,
    (*e*) 11000100000, (*f*) 110111011010, (*g*) 10000111010011.
3. (*a*) 11011010, (*b*) 10100011101, (*c*) 101011100110,
    (*d*) 1001001001011, (*e*) 1111110110101, (*f*) 10000100110000.
4. (*a*) 281, (*b*) 164, (*c*) 45, (*d*) 599, (*e*) 5261, (*f*) 2566.
5. (*a*) 355, (*b*) 21330, (*c*) 2576, (*d*) 1200100.
6. (*a*) 534, (*b*) 1522, (*c*) 4612, (*d*) 6, (*e*) 7, (*f*) 71.
7. (*a*) 453, (*b*) 1002, (*c*) 2101.
8. (*a*) 0·5625, (*b*) $0·9\dot{0}$, (*c*) $0·\dot{3}$, (*d*) $0·\dot{4}$, (*e*) $0·\dot{8}46153\dot{3}$, (*f*) 0·6875, (*g*) $0·\dot{7}$.

Chapter 5

Exercise 16   *page* 79
4. (*a*) 3, (*b*) 15, (*c*) 4, (*d*) 2, (*e*) 0, (*f*) 3, (*g*) 3, (*h*) 4, (*i*) 0, (*j*) 3.

Exercise 17 *page* 80
1. $4n+1$ (or it could be $4n-3$ or any other suitable form; but they are all fundamentally the same). 2. $7n+3$.
3. $8n+2$. 4. $5n+2$. 5. $23, 40, 57$. 7. Yes: $5n+4$.

Exercise 18 *page* 84
1. (a) 1, 2, 3, 4, 5, 6, 10, 12, 15, 20, 30, 60, (b) 1, 2, 3, 4, 6, 7, 12, 14, 21, 28, 42, 84, (c) 1, 5, 7, 11, 35, 55, 77, 385.
2. (a) 2, 3, 5, 13, (b) 11, 13, 17, (c) 3, 7, 13, 137.
3. (a) No, (b) Yes, (c) No, (d) No.
4. (a) div 11, (b) div 13, (c) None.
5. 3, 5, 7, 15. 6. 3, 7, 13.

Chapter 6

Exercise 19 *page* 88
1. (a) Yes, (b) Yes, (c) No, (d) No, (e) Yes. 2. Yes, yes, no, yes, no.
3. (a) Yes, (b) No. 4. ϕ, {1}, {3}, {5}, {7}, {1, 3}, {1, 5}, {1, 7}, {3, 5}, {3, 7}, {5, 7}, {1, 3, 5}, {1, 3, 7}, {1, 5, 7}, {3, 5, 7}, {1, 3, 5, 7}; 16 subsets. 5. {1, 2, 3, 5, 7, 11, 13}; 128.
6. (a) Yes, (b) Yes, (c) Yes, (d) No. 7. Yes. 8. 47; 257.
9. 256.

Exercise 20 *page* 91
2. (a) Queen of diamonds, (b) ϕ, {Q ♦ }, {Q ♥ }, {Q ♦ , Q ♥}.
3. (a) Yes, (b) Yes. 5. (a) Yes, (b) No, (c) Yes, (d) Yes, (e) No, (f) Yes; Rewording: $A =$ {Squares}, etc.
6. Fallacy: the polygon may have more than 4 angles.
8. $\mathscr{E} =$ {egg-laying animals}.

Exercise 21 *page* 94
1. {3, 7, 10}, {1, 3, 7}, {2, 3, 7}. 2. {3, 7}.
3. (a) {2, 5, 7, 8}, (b) {1, 3, 5, 7}, (c) {4}, (d) ϕ. 4. Yes.
5. {Squares}.
6. This is a trick question: Sydney might be a boy or a girl but with this spelling is more probably the latter; *ergo*, we do not know whether a bicycle can be ridden.
7. The circumcentre, i.e. the centre of the circumscribed circle, of triangle ABC.

Exercise 22   *page* 98
2. ϕ.   5. {1, 6, 9}, {2, 4, 6}, {1, 2, 3, 4, 6, 7, 8, 9},
    {1, 2, 3, 4, 6, 7, 8, 9}.   9. *A.*   10. (*a*) True, (*b*) False,
    (*c*) False.

Chapter 7

Exercise 23   *page* 106
Questions 1 (*a*), 2 (*a*), 3(*a*) are fundamentally the same as 1 (*b*), 2 (*b*),
3 (*b*) respectively, but written a slightly different form. The answers
are therefore given in corresponding notation.
1. (*a*) $x > 8$, (*b*) $S = \{x : x > 8\}$.   2. (*a*) $x > \frac{4}{5}$, (*b*) $S = \{x : x > \frac{4}{5}\}$.
3. (*a*) $x \geqslant -\frac{1}{5}$, (*b*) $S = \{x : x \geqslant -\frac{1}{5}\}$.   4. $-\frac{3}{2} < x < \frac{3}{2}$.
5. $-2 \leqslant x \leqslant 4$.
7. $-4 \leqslant x < -2\frac{1}{2}$ and $3 < x \leqslant 4$;
    $S = \{x : -4 \leqslant x < -2\frac{1}{2}\} \cup \{x : 3 < x \leqslant 4\}$.

Exercise 24   *page* 111
1. The part of the *x*, *y* plane lying above the line $y = x + 2$ and
    including the line itself.
2. The part of the *x*, *y* plane lying to the left of the line $y = -2x$
    and not including the line itself.
3. The part of the *x*, *y* plane above the line $y > -\frac{5}{4}$ and not includ-
    ing the line itself.

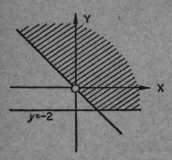

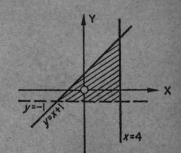

4. The shaded area illustrated.   5. The shaded area illustrated.

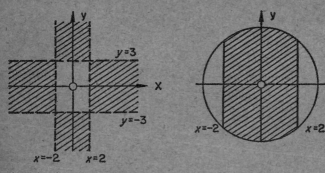

6. The shaded area illustrated.   7. The shaded area illustrated.

Exercise 25   *page* 114
1. $(1, -2), (1, 2), (2, -2), (2, 2), (1, 0), (2, 0), (3, 0)$.
2. $(0, 0), (-2, 1), (-1, 1), (0, 1), (1, 1), (2, 1), (-2, 2), (-1, 2), (0, 2),$
    $(1, 2), (2, 2)$.   3. $(1, 1), (2, 0)$.

Chapter 8

Exercise 26   *page* 117
1. (a) A, (b) A.   5. (a) $A \cap (B \cup C)$, (b) $A \cup (B \cap C)$.

Exercise 27   *page* 119
1. (a) $7; 2^4 . 3 . 5 . 7 . 11 = 18480$, (b) $6; 17136$.
2. (a) $z; x^3 y^3 z^2$, (b) $2c; 240a^3 b^2 c^2$.
3. $2(a-b); 12ab(a-b)$.   4. $256; 1536$.

Exercise 28   *page* 122
3. (a) $\{2, 4, 6, 8\}$, (b) $\{1, 5, 8\}$, (c) $\{8\}$, (d) $\{1, 2, 4, 5, 6, 8\}$, (e) $\{1, 5\}$,
    (f) $\{8\}$, (g) $\{1, 2, 4, 5, 6, 8\}$.

6. $\{a, b, e, f\}$, which is, of course, C.   7. 18 going only to the
    match.   8. $22; 3; 53$.

Exercise 29   *page* 127
1. $A+B$.   2. A'.   3. 0.   4. 1.   5. B.
6. A.   7. A.   8. $A+B$.   9. $B+AC$.   10. 0.

11. $AB+AC$.   12. $A'B$.   13. $A'B'+AB$.   14. $A+BC$.
15. Page 112; $A'B' = (A+B)'$.

Exercise 30   *page* 129
2. $\{1, 2, 5, 6\}$.

Chapter 9

Exercise 31   *page* 135
1. Yes.   2. Yes.

3. Doubtful: it depends whether or not *all* viewers become lazy.

4. No: it may be commonsense but it is not logic.

5. Strictly, the conclusion should be "Some numbers are multiples of 12", for 2 is included in 4.

6. No. The only logical conclusion would be that each businessman has a chauffeur who drives the car away.

7. No. It is a play on words.

8. Some boys spend pocket-money on cigarettes.

9. Some women cannot afford to buy new hats.

10. N is a multiple of 42.

11. No solution because the words *up* and *down* are used in two different senses at the same time.

Exercise 32   *page* 141
7. The statement is not true.

Exercise 33   *page* 142
3. $(a_N a)_N (b_N b)$.   4. $(a_N b)_N (a_N b)$.
5. $(a_N a)_N (b_N c)$.   6. $[(a_N a)_N (b_N b)]_N [(a_N a)_N (b_N b)]$.
7. $[a_N \{(b_N b)_N (c_N c)\}]_N [a_N \{(b_N b)_N (c_N c)\}]$.   8. $(a_N a)_N (a_N a)$.

Chapter 10

Exercise 34   *page* 149
1. $ab+cd$.   2. $a+b+c$.   3. $[(a+b)c+d]e$.
4. $(a+c)b+(e+f)d$.   5. No; Yes; Yes; No.
6. No; Yes; No; Yes.

In (i)–(iv), A, B, C, D, E, F are simple switches.

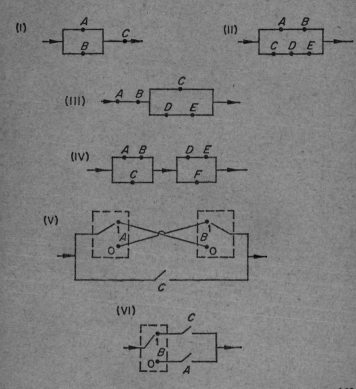

(I)

(II)

(III)

(IV)

(V)

(VI)

Exercise 35 page 163

2. 31 cannot be obtained for inputs of 15 and 16 respectively would be needed; the number 15 is 1111_2 (4 digits) but 16 is 10000_2 (5 digits) and the latter is not possible on the instrument shown.

3. (a) Fig. 110; (b)

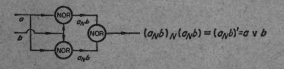

$$(a_N b)_N (a_N b) = (a_N b)' = a \lor b$$

Chapter 11

1. Examples only are shown to indicate the shape of the matrices.

(a) $\begin{pmatrix} 2 & -5 \\ 0 & 1 \end{pmatrix}$,   (b) $(2 \quad 4 \quad -1)$,   (c) $\begin{pmatrix} 3 \\ 2 \end{pmatrix}$,

(d) $\begin{pmatrix} a & b & c \\ -a & d & 0 \\ e & b & f \end{pmatrix}$,   (e) $\begin{pmatrix} 0 & 4 & 2 \\ 3 & -5 & 6 \end{pmatrix}$,   (f) $\begin{pmatrix} p & q \\ -p & q \\ 0 & 1 \end{pmatrix}$.

2. 5 rows, 3 columns.
3. (a) Yes, (b) No, (c) Yes, (d) No, (e) Yes, (f) Yes, (g) Yes.

(a) $\begin{pmatrix} 2 & 5 \\ 0 & 2 \end{pmatrix}$, (c) $(0 \quad 2 \quad -5)$, (e) $\begin{pmatrix} -1 \\ -1 \end{pmatrix}$

(f) $\begin{pmatrix} 6 & 4 \\ 10 & 0 \\ 0 & 3 \end{pmatrix}$, (g) $\begin{pmatrix} 2 & 2 & 9 \\ -8 & 2 & 8 \\ 7 & 0 & 4 \end{pmatrix}$.

4. (a) $x = -2$, (b) $x = \frac{2}{3}$, $y = \pm 2i$, (c) $x = -1$, $y = 0$,
   (d) $x = -2$, $y = 3$, (e) $x = 3$ or 4.

5. $x = \pm\frac{3}{2}$; y cannot be determined.

1. $\begin{pmatrix} 5 & 10 \\ 15 & -20 \\ 0 & 25 \end{pmatrix}$ 2. $\begin{pmatrix} 4 & 0 & -16 \\ -8 & -4 & 12 \\ 8 & -20 & 0 \end{pmatrix}$ 3. $\begin{pmatrix} 0 & 2 & 12 \\ 0 & 0 & 0 \end{pmatrix}$

4. $\begin{pmatrix} 0 & 0 & 0 \\ 0 & 0 & 5 \\ 1 & 0 & 3 \end{pmatrix}$.   5. $x = -\frac{1}{3}$,
                                $y = 2$.

6. (a) $17\begin{pmatrix} 1 & 3 \\ -2 & 0 \end{pmatrix}$, (b) $7\begin{pmatrix} 13 & 9 & -8 \\ 7 & 0 & -14 \end{pmatrix}$,

1. (a) Yes, (b) Yes, (c) Yes, (d) Yes, (e) No, (f) Yes, (g) Yes, (h) Yes,
   (i) Yes.

(a) (-15), (b) (-4), (c) $(6 \quad 13)$, (d) $\begin{pmatrix} 8 & 6 \\ 4 & 3 \end{pmatrix}$, (f) (11),

$(g) \begin{pmatrix} 10 & 8 \\ 0 & 0 \\ -15 & -12 \end{pmatrix}$, $(h) \begin{pmatrix} 12 \\ 3 \\ 6 \end{pmatrix}$, $(i) \begin{pmatrix} 2 & 10 \\ 14 & 19 \\ -3 & 2 \end{pmatrix}$

2.
$\mathbf{AB} = (8)$, $\mathbf{BA} = \begin{pmatrix} 4 & 2 & 0 \\ 8 & 4 & 0 \\ 6 & 3 & 0 \end{pmatrix}$.

3. $\mathbf{PQ} = \begin{pmatrix} 6 \\ -5 \end{pmatrix}$; $\mathbf{QP}$ has no meaning. 4. $\begin{pmatrix} 6 & -3 & 41 \\ -1 & -3 & 6 \\ 19 & 11 & -7 \end{pmatrix}$.

5. $x = 12, y = -10$. 6. $\begin{pmatrix} 0 & 12 & 2 \\ 1 & 19 & 0 \end{pmatrix}$.

7. $a_{32} = 11, a_{21} = -2$.

Exercise 39 *page* 180
1. $x = 2, y = 0$. 2. $m = \frac{2}{3}, n = -\frac{3}{2}$. 3. $x = -\frac{3}{5}, y = \frac{13}{10}$.

Exercise 40 *page* 183
1. $\mathbf{A}^2 - \mathbf{A} - 2\mathbf{I} = \phi$. 2. $\mathbf{A}^2 - 2a\mathbf{A} + (a^2 + b^2)\mathbf{I} = \phi$.
3. $\mathbf{A}^2 + 7\mathbf{I} = \phi$. 4. $\mathbf{A}^2 - 2(c + d)\mathbf{A} = \phi$.
5. $\mathbf{P}$, No; $\mathbf{Q}$, Yes; $\Delta(\text{for } \mathbf{Q}) = -8$.

6. $\begin{pmatrix} a^3 + 2a^2b - ab^2 & a^3 + ab^2 \\ a^2b + b^3 & a^2b - 2ab^2 - b^3 \end{pmatrix}$ 7. $49 \begin{pmatrix} -1 & -3 \\ 6 & 2 \end{pmatrix}$.

8. (a) $\mathbf{X} = \frac{1}{5} \begin{pmatrix} 13 \\ 1 \end{pmatrix} \Rightarrow x = \frac{13}{5}, y = \frac{1}{5}$; (b) $x = \frac{2c}{2a-b}, y = \frac{-c}{2a-b}$.

Exercise 41 *page* 186
1. 73% opposed, 27% in favour; New situation: 74·7% opposed, 25·3% in favour. 2. 12·25% in favour, 5% undecided, 82·75% opposed. 3. 35·51% in favour of Shrinko, 35·87% in favour of Rottaway, 28·62% in favour of Skratch.

Chapter 12
Exercise 42 *page* 191
1. $\frac{1}{2}$ (the earlier results have no bearing in this case).
2. $\frac{1}{3}$ (theoretical). 3. Probability of Smith having prime number $\frac{2}{3}$; probability of Jones having prime number $\frac{5}{8}$; hence Smith should win.

4. (a) $\frac{3}{26}$, (b) $\frac{5}{13}$.   5. (a) $\frac{11}{50}$, (b) $\frac{3}{50}$, (c) $\frac{12}{25}$.

6. $\frac{3}{4}$, 3:1; (a) 8, (b) 5. (This is the worst possible result for, of the first 4 draws, the most unfavourable situation is 2W, 2R and then the next must be W or R.)

Exercise 43 page 194

1. (a) $\frac{1}{6}$ (b) $\frac{5}{36}$.   2. 67:5; $\frac{1}{36}$.   3. $\frac{5}{6}$.   4. 5:3.   5. 9:2.

Exercise 44 page 196

1. $\frac{25}{36}$; $\frac{11}{36}$.   2. $\frac{1}{9}$.   3. John; 125:91.

Exercise 45 page 197

1. (a) $\frac{1}{1326}$, (b) $\frac{1}{17}$.   2. $\frac{1}{22100}$.   3. $\frac{4}{7}$.

5. $\left(\frac{211}{243}\right)^3$.   6. $\frac{3}{8}$.   7. 50:49.

8. $\frac{32}{63}$.   9. $p_1(1-p_2):(p_1+p_2-2p_1p_2)$.

Exercise 46 page 203

1. $\frac{5}{16}$; Yes.   2. (a) 0·262, (b) 0·246.   3. 6560:1.

4. $1-\left(\frac{35}{36}\right)^{24} \doteqdot 0·490$ (using log. tables); losing bet.

6. (a) $^4C_2(\frac{1}{6})^2(\frac{5}{6})^2 = \frac{25}{216}$,

    (b) $^4C_2(\frac{1}{6})^2(\frac{5}{6})^2 + {}^4C_3(\frac{1}{6})^3(\frac{5}{6}) + {}^4C_4(\frac{1}{6})^4 = \frac{19}{144}$.

7. $\frac{1}{6}\times 1 + \frac{5}{11}\times\frac{5}{6} = \frac{6}{11}$.

Exercise 47 page 209

1. $\frac{3}{640}$.   2. $\frac{5}{34}$.   3. $\frac{1}{126}$.

TEACH YOURSELF BOOKS

BUSINESS, PROFESSIONAL, COMPUTERS AND OFFICE PRACTICE

Advertising
E. McGregor — 50p

Banking
J. P. Parker &
B. D. W. Cox — 50p

Book-keeping
D. Cousins — 50p

Catering and Hotel Operations
V. G. Winslet &
S. Blundell — 40p

Commerce
Ronald Warson — 60p

Commercial Arithmetic
J. H. Harvey — 40p

Commercial Correspondence
F. Addington-Symonds — 50p

Costing
D. Cousins — 40p

Cybernetics
F. H. George — 40p

Dutton Speedwords
R. G. J. Dutton — 40p

Economics
S. E. Thomas — 40p

Electronic Computers
F. L. Westwater &
D. H. Joyce — 40p

Exporting
D. F. Taylor &
E. A. Rutland — 50p

Insurance
H. A. L. Cockerell — 50p

Investment
W. L. B. Fairweather — 40p

Management Accounting
B. Murphy — 50p

Office Practice
J. Shaw — 50p

Operational Research
M. S. Makower &
E. Williamson — 60p

O & M
R. G. Breadmore — 50p

Public Relations
H. Lloyd — 50p

Salesmanship
S. A. Williams — 40p

Secretarial Practice
Pitmans College — 50p

Shorthand
Pitmans College — 40p

Typewriting
Pitmans College — 40p

MATHEMATICS, SCIENCE AND TECHNICAL

Algebra
P. Abbott — 50p

Arithmetic: Decimalized and Metricated
L. C. Pascoe — 50p

Biochemistry
P. H. Jellinck — 40p

Biology
J. R. Hall — 75p

Biology, Human
D. Taverner — 50p

Botany
J. H. Elliott — 40p

Calculus
P. Abbott — 50p

Calculus, Further
F. L. Westwater — 50p

Concreting, Practical
A. E. Peatfield — 60p

Dynamics
C. G. Lambe — 40p

Electricity
C. W. Wilman — 40p

Electronics
W. P. Jolly — 50p

Electronics, Industrial
C. W. Eggleton — 50p

Engineering Science
C. B. Day & V. A. Jones — 50p

Farming
James Merridew — 60p

Genetics
R. J. Berry — 50p

Geology
A. Raistrick — 60p

Geometry
P. Abbott — 50p

Human Anatomy and Physiology
David Le Vay — 95p

Inorganic Chemistry
R. E. Morcom — 60p

Mathematics
J. Davidson — 40p

Mathematics, New
L. C. Pascoe — 50p

Mechanical Engineering
A. E. Peatfield
Vol 1 Hand Tools — 40p

Vol 2 Engineering Components — 50p

Vol 3 Workshop Practice — 50p

Mechanics
P. Abbott — 60p

Metrication
J. Peach — 40p

Microbiology
G. D. Wasley & R. W. Warner — 60p

Nutrition
M. Pyke — 40p

Organic Chemistry
K. Rockett — 60p

Perspective Drawing
H. F. Hollis — 40p

Physics
D. Bryant — 50p

Plumbing
J. H. Innes — 50p

Seamanship
T. F. Wickham &
N. Hefford — 50p

Slide Rule
B. Snodgrass — 40p

Statistics
R. Goodman — 50p

Trigonometry
P. Abbott — 40p

The Weather
Graham Sutton — 60p

Welding
C. Bainbridge — 50p

Zoology
T. M. Savory — 50p

ARTS AND GENERAL

Analytical Psychology
D. Cox — 50p

Architecture in England
T. W. West — 50p

Art, Study
D. Talbot Rice — 50p

**Asesssing Intellectual
Ability**
B. A. Akhurst — 40p

Astrology
Jeff Mayo — 50p

Astronomy
D. S. Evans — 50p

Codes and Ciphers
F. Higenbottam — 50p

Creative Writing
Victor Jones — 50p

Drawing
R. Smith — 50p

Drawing and Painting
S. A. Knight — 40p

Efficient Reading
C. Mares — 40p

**Encyclopaedia of Dates and
Events** — £1.75
L. C. Pascoe, A. J. Lee
E. S. Jenkins, H. V. Ralph

Ethics
A. C. Ewing — 40p

**Etiquette and Good
Manners**
W. S. Norman — 40p

Express Yourself
R. W. Jepson — 40p

Handwriting
J. le F. Dumpleton — 40p

Heraldry and Genealogy
Leslie Pine — 75p

History of England
W. McElwee — 50p

Law
J. Leigh Mellor 75p

Letter Writer
G. S. Humphreys 50p

Logic
A. A. Luce 50p

Music
C. King Palmer 40p

Music, Compose
C. King Palmer 60p

Music, History of
A. Percival 60p

Navigation
A. C. Gardner 40p

Orchestration
C. King Palmer 50p

Organ, The
F. Routh 50p

Painting for Pleasure
R. O. Dunlop 40p

Personal Efficiency
F. Addington-Symonds 50p

Philosophy
C. E. M. Joad 50p

Philosophy, History of
J. Lewis 50p

Philosophy of Religion
M. D. Lewis 60p

Piano, The
C. King Palmer 50p

Piano, Jazz
Eddie Harvey 95p

Poetry
R. Skelton 40p

Political Thought
C. L. Wayper &
C. W. Parkin 50p

Poultry Keeping, Modern
J. Portsmouth 60p

Public Speaking
P. Westland 40p

Radio Servicing
L. Butterworth 50p

Semantics
F. H. George 40p

Singing
C. R. Thorpe 40p

Sociology
J. H. Abraham 50p

Speaker and Debater 40p

Spelling
K. Baron 50p

Study
G. G. Neill Wright 50p

Understanding Statistics
Martin Leonard 60p

Violin, The
T. Rowland-Entwistle 50p

Yoga
J. Hewitt 40p

Zen
C. Humphreys 50p